Lecture Notes on Coastal and Estuarine Studies

Managing Editors:
Richard T. Barber   Christopher N. K. Mooers
Malcolm J. Bowman   Bernt Zeitzschel

9

Osmoregulation in Estuarine and Marine Animals

Proceedings of the Invited Lectures to a Symposium
Organized within the 5th Conference of the European Society
for Comparative Physiology and Biochemistry –
Taormina, Sicily, Italy, September 5 – 8, 1983

Edited by A. Pequeux, R. Gilles and L. Bolis

Springer-Verlag
Berlin Heidelberg New York Tokyo 1984

QL
121
.O86
1984

ISBN 3-540-13353-4 Springer-Verlag Berlin Heidelberg New York Tokyo
ISBN 0-387-13353-4 Springer-Verlag New York Heidelberg Berlin Tokyo

EUROPEAN SOCIETY FOR COMPARATIVE PHYSIOLOGY AND BIOCHEMISTRY

5th Conference - Taormina, Sicily - Italy , September 5-8, 1983

*This volume gathers the proceedings of the invited lectures of
the symposium on Osmoregulation in Estuarine and Marine Animals*

Conference general theme and symposia

Physiological and Biochemical Aspects of Marine Biology

Symposia : 1) Toxins and drugs of marine animals
2) Responses of marine animals to pollutants
3) Marine phytoplankton and productivity
4) Osmoregulation in estuarine and marine animals

Conference Organization

General organizers

L.BOLIS and R.GILLES

Messina, Italy/Liège, Belgium

Symposium scientific organizers

for Osmoregulation in Estuarine and Marine Animals.
A.PEQUEUX and R.GILLES
Liège - Belgium

Local organizers

G.STAGNO d'ALCANTRES, S.GENOVESE, G.CUZZOCREA, F.FARANDA,
A.CAMBRIA

Messina, Italy

Local secretariat

A.SALLEO and P.CANCIGLIA

Messina, Italy

Conference under the patronage of

The University of Messina , Italy

The Fidia Research Laboratories, Italy

The European Society for Comparative Physiology and
Biochemistry

CONTENTS

PART I

PHYSIOLOGICAL AND ULTRASTRUCTURAL ASPECTS OF "SALT-TRANSPORTING TISSUES" STUDIES

PART II

Biophysical and Biochemical Aspects of "Salt-Transporting Tissues" Studies

LIST OF AUTHORS AND CONTRIBUTORS

T.J.BRADLEY : Dept. of Developmental and Cell Biology,
School of Biological Sciences
University of California,
Irvine, CA 92717
U.S.A.

W.A.DUNSON : Dept. of Biology, 208, Mueller Laboratory,
The Pennsylvania State University, University Park,
PA 16802, U.S.A.

S.A.ERNST : Dept. of Anatomy and Cell Biology,
University of Michigan,
School of Medicine, Ann Arbor,
Michigan 48109
U.S.A.

B. FOSSAT : Laboratoire de Physiologie Cellulaire et Comparée,
Faculté des Sciences et des Techniques,
Parc Valrose, 06034 NICE Cedex
FRANCE

R.GILLES : Laboratoire de Physiologie Animale,
Université de Liège, 22, quai Van Beneden
B-4020 Liège
BELGIQUE

B.L.GUPTA : Dept. of Zoology, University of Cambridge,
Downing street, Cambridge CB2 3EJ
ENGLAND

S.R.HOOTMAN : Dept. of Physiology, University of California
School of Medicine, San Francisco,
CA 94143
U.S.A.

W.HUMBERT : Laboratoire de Zoologie et d'Embryologie Expérimen-
tale, A.I.CNRS 033669, Université Louis Pasteur,
12, rue de l'Université, 67000 Strasbourg, FRANCE

R.KIRSCH : Laboratoire de Zoologie et d'Embryologie Expérimen-
tale, A.I. CNRS 033669, Université Louis Pasteur,
12, rue de l'Université, 67000 Strasbourg, FRANCE

B.LAHLOU : Laboratoire de Physiologie Cellulaire et Comparée,
Faculté des Sciences et des Techniques, Parc Valrose,
06034 Nice Cedex, FRANCE

C.LERAY : Laboratoire de Physiologie Comparée des Régulations,
CNRS, B.P. 20CR, 67037 Strasbourg Cedex, FRANCE.

B.G. MUNCK : Dept. of Medical Physiology A, The Panum Institute,
University of Copenhagne , DK 1870 Copenhagen.
DENMARK

A.J.R. PEQUEUX : Laboratoire de Physiologie Animale,
Université de Liège, 22, Quai Van Beneden,
B-4020 Liège
BELGIQUE

J.E. PHILLIPS : Dept. of Zoology, University of British Columbia,
Vancouver, B.C. V6P 1A9 CANADA

G.E. RICE : Dept. of Physiology, Monash University,
Victoria, AUSTRALIA

J.L. RODEAU : Laboratoire de Physiologie Respiratoire,
CNRS, 23, rue du Loess, Strasbourg
FRANCE

E.SKADHAUGE : Dept. of Veterinary Physiology and Biochemistry
The Royal Veterinary and Agricultural University,
Bülowsvej 13, DK 1870 Copenhagen V
DENMARK

K.STRANGE : Dept. of Zoology, University of British Columbia
Vancouver, BC V6P 1A9
CANADA

Present address : Lab. of Kidney and Electrolyte
Metabolism, National Institutes of Health,
Bethesda, MD 20205, USA

D.W.TOWLE : Dept. of Biology, University of Richmond,
Richmond, Virginia 23173
U.S.A.

J.A.ZADUNAISKY : Dept. of Physiology and Biophysics and Ophthalmo-
logy and Visual Sciences
New York University Medical Center,
School of Medicine, 550 First Avenue,
New York, N.Y. 10016
U.S.A.

I.ZERBST-BOROFFKA : Institut für Tierphysiologie,
Fachbereich Biologie (FB23), Freie Universität
Berlin, 34 Grunewaldstrasse, D-1000 Berlin 41
R.F.A.

PREFACE

A wealth of information on osmotic and ionic regulation in
Estuarine and Marine Animals has been accumulated over the past
decades.
Beyond early studies of whole-animal responses to changes in environmental salinities, efforts have been made later on to identify,
to localize and to characterize the organs and structures responsible
for the control of the characteristics of the cell's environmental
fluid. When considering the problem of cell volume control in animals
facing media of fluctuating salinities, we are indeed dealing with
two different categories of mechanisms. A first one is concerned
with the control of the osmolality of the intracellular fluid, hence
with the processes directly implicated in the maintenance of cell
volume and shape. They have been extensively described in several
recent review papers.
The second category includes the processes controlling the characteristics of the cell's environmental fluid in order to minimize the
amplitude of the osmotic shocks the cells may have to cope with upon
acclimation to media of changed salinities. They are localized in
particular organs and structures : the so-called *"salt-transporting"*
epithelia. Up to now, most of the studies on salt-transporting epithelia in estuarine and marine animals used the black box approach, so
that little or sometimes nothing is still known on the physiological,
the biochemical and the biophysical basis of the transporting mechanisms as well as on the structure-function relationships. With respect
to the mechanisms concerned, recent works with *in vitro* preparations
of isolated salt-transporting organs have allowed kinetical characterization of the transport processes. Combined with electron microscopy
investigations, they should result in the opening of the transport
black box and in a better integration of its structural organization
with its function. As a further step, subcellular localization of enzymes related to transport could also be integrated with physiological
and biochemical data to yield or functional model of passive and active movements of ions.

This volume plans to summarize, from a comparative point of view, the progress that has been made in understanding how the control of the body fluids is achieved in a wide range of estuarine and marine organisms. Our goal is to center essentially on the salt-transporting tissues from integuments and nephridia of molluscs and annelids to the salt gland of reptiles and birds, emphasizing the prominent part they play in the regulation of body fluids salt content. A first part will thus be centered on a comparative review of the essential ultrastructural and physiological features of the salt transporting epithelia at work in various estuarine and marine animal groups.

The second part of this volume will consider , from a more general point of view, some biophysical and biochemical characteristics of the transport mechanisms involved in the osmo and ionoregulation achieved by these tissues and organs.

Finally, in an attempt to correlate structure and function, models of salt and water flow across epithelia will be proposed.

It is clear that more basic research is needed to completely and satisfactorily bridge the gap between structure and function but concerted studies in ecology, morphology, physiology and biochemistry as in the context of the meetings of the European Society for Comparative Physiology and Biochemistry should, by no doubt, be a right way to hit that target.

Liège, December 1983.   A.PEQUEUX.

P A R T I

Physiological and Ultrastructural Aspects of
"salt-transporting tissues" Studies.

Homeostatic function of integuments and Nephridia in Annelids

I. ZERBST-BOROFFKA

Annelids are abundant in marine, estuarine, and fresh water environments. Especially in the estuaries -with strongly changing osmolarities of the environment- annelids need behavioural and physiological
mechanisms to minimize changes of their body volume and osmotic concentration. This field has been reviewed comprehensively by Oglesby,
1978. The review, presented here, deals especially with recent studies of the physiological mechanisms of salt and water balance. It
includes turnover studies, structure and function of the integument,
structure and function of the nephridia, a possible feed-back control
system and efferent control mechanisms. The problem of intracellular
volume regulation is neglected.

I. Physiological responses to changes of medium salinity

The response of extracellular fluid concentration to a change of external salinities is different in several species of annelids. Hyperosmotic conforming in higher salinities and regulating in lower salinities is the most widespread kind of response in euryhaline species
(fig.1) but it holds too for the more stenohaline species (fig.1). In
each species the level of the blood concentrations is individually
regulated and the salinity range where conforming and regulating
occur (fig.1) is characteristic, too.

A further problem for soft bodied animals living in altered medium
concentrations is the water balance. Oglesby, 1975 and Machin, 1975
both presented a theoretical analysis of the water balance of annelids in order to elucidate the ability and capacity of water regulation. These studies are based on the determination of the relative
water content (or water gain, resp.) compared with the pure osmotic
water change of the animal adapted to different salinities. Certainly
this treatment gives a better information of the ability of volume
regulation than the mere determination of the weight changes as used

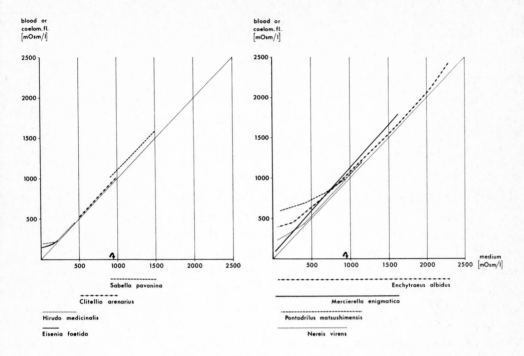

Fig.1. Steady-state blood concentrations of some representative annelids acclimated to different salinities. Sabella: Koechlin, 1975; Clitellio: Ferraris and Schmidt-Nielsen, 1982; Hirudo: Boroffka, 1968;
Eisenia: Takeuchi, 1980c; Enchytraeus: Schöne, 1971; Mercierella:
Skaer, 1974; Pontodrilus: Takeuchi, 1980d; Nereis: Oglesby et al.,
1982, Quinn and Bashor, 1982. (arrows: 100% sea water).

by Reynoldson and Davies, 1980. Oglesby, 1981 pointed out that an
approach to analyse the ability of water regulation should include
the change of dry weight as discussed in some species (Freeman and
Shuttleworth, 1977a,b, Linton et al., 1982). However, the determination of relative water content in different salinities may be useful
for interspecies comparison. Different types of volume regulation are
established, but the capacity to regulate the body volume is not consistent with the capacity to regulate the osmotic concentration. Hyperosmotic conformers as Clitellio arenarius and Mercierella enigmatica e.g. are entirely volume regulators (Ferraris and Schmidt-Nielsen, 1982; Skaer, 1974). Osmoregulation and volume regulation are
different processes. They are certainly closely interrelated but un-
derlie different controlling mechanisms (cf. Gauer, 1978; Oglesby,
1975; Zerbst-Boroffka, 1978). Volume changes are not at all 'by-pro-
ducts of osmotic regulation' as stated by Linton et al., 1982.

What are the mechanisms responsible for these different types of physiological response and which organs are involved?

II. Integumental properties

The estimation of turnover rates of solutes and water is one of the
common experimental approaches to investigate the mechanisms of solute and water balance. In annelids acclimated to different salinities
the influx and efflux of water, some ions, the uptake of organic substances from diluted media and the output of ammonia were measured
(cf. Oglesby, 1978; Ahearn and Gomme, 1975; Siebers et al., 1976,
1977, 1978; Mangum et al., 1978; Richards and Arme, 1979, 1980). The
uptake of organic substances was discussed as important to nutrition
but especially amino acids are essential for volume- and osmoregulation (Freeman and Shuttleworth, 1977c; Takeuchi, 1980b,d; König et
al., 1981; Carley et al., 1983). Some transport characteristics such
as saturation kinetics, independency of other ions and molecules, and
inhibition or stimulation by drugs could be determined by applying
isotopes. Additionally in some species the electrical trans body wall
potential could be measured and the transport process could be identified as active or passive (ref. in Oglesby, 1978; Mangum et al.,
1978).

Those studies in intact animals include the gut function. In Lumbricus terrestris Cornell, 1982 demonstrated a sodium conserving function
of the gut. Studies of drinking rates in annelids acclimated to different salinities are missing, only the experiments done by Ahearn and
Gomme, 1975 exclude drinking in Nereis diversicolor adapted to 50%
sea water. Since the oral uptake of solutes and water is assumed to
be unimportant in annelids the measured uptake rates are generally
interpreted as an achievement of the integument.

Studies with isolated body wall preparations exclude this problem but
bear all the complications of isolated preparations (removal from
control mechanisms and circulation). In Nereis diversicolor, N. succinea and Lumbricus terrestris acclimated to salinities from 50% sea
water down to nearly tap water salt and water fluxes were studied
both with intact animals and with isolated body walls (ref. in Oglesby, 1978; Cornell, 1982). The results show similar trends: with
higher medium dilution the apparent integumental permeability to
salts declines (limiting the salt loss), the apparent integumental
permeability to water declines, too (limiting the osmotic water in-
flow), the active salt uptake increases (compensating the passive

salt loss).

The site of integumental water and ion transport should be more speci-
fied. With the ecological (genetic) adaptation to various environments
many annelids developed differentiations of the body surface: e.g.
gills in most of the polychaeta and many hirudinea. Such body appen-
dages are well qualified to be the site not only of respiration but
also of ion and water transport. The control of the circulation in
some gills or gill segments or nervous/humoral stimulation and inhibition of solute pumps offer a functional advantage to change transport rates.

In many polychaeta the numerous parapodia are equally supplied with
an intense network of blood capillaries. In Nereis succinea a separation of respiratory and osmotic exchange surface is indicated by the
investigation of the sodium/potassium activated ATPase in various
tissues after acclimatization to different salinities (Mangum et al.,
1980; Mangum, 1982): In salinities of 100% sea water -where no net
sodium/chloride transport has been established- no significant transport enzyme activity could be detected in the body except a very
small activity in the upper notopodial ligule of the parapodia in the
anterior body region. When acclimated to low salinities -where active
sodium/chloride uptake has been established (Doneen and Clark, 1974)-
the transport enzyme activity was detected only in tissue preparations
containing parapodial tissue, and it was highest in the upper notopodial ligule of anterior parapodia. That means, that in Nereis succinea
the sites of active salt uptake are probably the upper notopodial li-
gules in the anterior body region.

More investigations should be performed to elucidate transport functions of special body surface areas.

The ultrastructure of the annelid integument was investigated in many
annelids, and was reviewed by Richards, 1978. Comparative EM studies
give no clear indications for transporting function of normal epider-
mal cells of the integument or the gills (Storch and Alberti, 1978).
Interestingly enough, many different kinds of gland cells were identified producing mucus of different composition (Hausmann, 1982). The
biological importance of mucus may be: defense against enemies or defense against dessication. In gastropods it was established recently
by Schlichter, 1982 that mucus behaves as a weak, negatively charged
ion exchanger, which is able to concentrate cations. If it functions
in annelids in the same way, the kinetics of cation uptake has to be
reinterpreted. The unstirred mucus layer may be regarded as a further

compartment between the medium and the integument. Mucus production in annelids is under nervous control (Lent, 1973). Such control system should be able to vary the mucus composition and the secretion rate. Thus the animal would be enabled to adapt the properties of the mucus compartment to the best function.

III. Nephridial function

All annelids have nephridia providing for the excretion of ions and water. The participation of the gut and rectum in the excretion of salt and water has not been investigated extensively. In the earthworm Lumbricus terrestris the rectal fluid is hypotonic (Cornell, 1982), suggesting some ion conserving function. Many important aspects e.g. the rectal volume output, however, were not investigated.

The discussion of the nephridial function in controlling salt/water balance is focused on the following problems:
A. Mechanisms of primary and final urine formation
   (including filtration, secretion and reabsorption processes)
B. The homeostatic effeciency
C. Control mechanisms of the nephridial function

Ultrafiltration as one of the processes of primary urine formation occurs in 2 morphologically different types of nephridia: - open nephridia with a nephrostome, - closed nephridia with terminal cells.

The big pair of thoracic nephridia in Sabella pavonina extensively studied by Koechlin, 1966, 1981, 1982 may be mentioned as an example of an open nephridium. The nephrostome cilia move coelomic fluid into the nephridial canal where the final urine is formed. The coelomic fluid is therefore regarded functionally as primary urine. In Sabella pavonina it is assumed to be filtrated at the periesophageal capillary network, since EM studies demonstrated a high permeability of the coelomic epithelium lining the capillaries (Koechlin, 1966). Podocytes as a morphological requirement for filtration of coelomic fluid from the blood have been demonstrated in the ventral vessel of the oligochaete Tubifex tubifex by ultrastructural studies (Peters, 1977).

In some polychaetes, e.g. Glycera unicornis and Nephthys hombergi the primary urine is probably filtrated from the coelomic fluid by the terminal cells (=solenocytes) called 'cyrtocytes' (Brandenburg, 1975; Kümmel, 1975; Wilson and Webster, 1974). This type of terminal cell occurs in some other animal groups, too. It is characterized by a weir system covered by the filter membrane and contains 1 flagellum. The primary urine flows into the efferent nephridial tubule.

It should be emphasized that at present the investigation of the problem of filtration including the structure involved, driving forces,
and resistances is focussed on molluscs and not on annelids.

In all hirudinea the connection between the nephrostome (or ciliated
funnel) and the following nephridial tubule has been lost during evolution. In the nephridium of the medical leech the process of the
formation of primary urine is secretion. The primary urine is secreted by the nephridial lobe cells into a system of canaliculi which
runs into the nephridial canal (Zerbst-Boroffka and Haupt, 1975).
Recently, micropuncture techniques and electrical potential measurements show that potassium and chloride are actively transported, while
sodium secretion is passive (Zerbst-Boroffka et al., 1982; Zerbst-
Boroffka and Wiederholt, 1982). Reabsorption of solutes in the central
canal provides for hypotonicity of the final urine (Zerbst-Boroffka,
1975; Wenning et al., 1980).

In Sabella pavonina it has been shown by Koechlin, 1981a,b, 1982,
that the nephridia are involved in controlling organic molecules.

The osmotic and ionic concentration of the final urine are important
indicators of the homeostatic function of the annelid nephridia. Some
species are able to produce only an iso-osmotic urine, e.g. Glycera
dibranchiata. In other species as in Nereis diversicolor the urine
becomes hypotonic in lower medium salinities if the blood concentration is regulated. If the nephridia are the only water excreting organs the ability to produce a hypotonic urine is obligatory for life
in diluted media or fresh water. Unfortunately, the urine concentrations have been measured only in a few annelids, probably because
sampling of urine is difficult in these small animals. Thus, a well
founded comparative work -necessary to draw general conclusions- is
lacking (Zerbst-Boroffka, 1977).

Direct measurements of urine flow rates (e.g. by applying catheters)
are only available for the medical leech. In fresh water we estimated
a urine flow of $5\mu l/hr \cdot cm^2$ (body surface). This is in agreement with
values of other fresh water animals. Acclimated to higher medium salinities the steady-state urine flow declined to nearly zero in 40%
sea water (Boroffka, 1968). Similar results were obtained in Nereis
by indirect methods (cf. Oglesby, 1978).

Sucking of blood represents a considerable osmotic stress for the
leech. The blood meal is strongly hypertonic, but the leech absorbs
remarkable volumes of salt solutions from the crop within some hours
after the meal and excretes it via the nephridia. Since the urine

concentration never becomes hypertonic an additional salt output has
to be assumed (Wenning et al., 1980).

IV. A possible feed-back control mechanism and efferent control factors

In our experiments on the leech we infused a hypertonic or a hypotonic salt solution into the crop and measured the urine concentration,
urine flow, blood concentration and blood volume. The results show
that the urine flow rises independently of the blood concentration
but correlates with the blood volume. Thus, the nephridial function
can not be controlled by the osmotic blood concentration alone. We
postulated a feed-back control system based on these results and
some further experiments. It could provide a homeostatic control mechanism of the water content. The factor to be controlled could be
the blood volume. Stretch receptors situated in the blood vessel wall
could measure the actual volume (or perhaps the blood pressure). The
information could be processed in the CNS. By hormonal or neuroendocrine mechanisms the urine flow and the integumental water output
could be stimulated or inhibited, thus correcting the blood volume
(Zerbst-Boroffka, 1978).

Certainly, such a model is simple and does not cover the complexity
of control mechanisms. However, it represents a further attempt to
understand the iso-osmotic volume regulation of osmoconformers.

While further studies of the afferent control mechanisms are lacking,
a few publications are devoted to the problem of efferent control mechanisms, namely neuroendocrine or hormonal mechanisms controlling
the integumental or nephridial function.

In some fresh water and estuarine annelid species histochemical investigation and brain ablation experiments were performed some of
which including replacement therapy (such as reimplantation of brain
or injection of brain homogenates). These experiments include determinations of water content, extra- and intracellular solutes and free
amino acid concentrations, turnover rates of salts and water. They
indicated that the brain takes part in water balance control and
perhaps in the control of free amino acids and solute balance (Carley, 1978; Nagabhushanam and Kulkarni, 1979; Takeuchi, 1980a,b,c;
Ferraris and Schmidt-Nielsen, 1982; Malecha, 1983).

Electrophysiological studies on efferent control of the integumental
function were performed only with the medical leech. In hirudinea
each segmental ganglion contains 2 giant neurons, the Retzius cells.

They send processes into the integument of the same segment. If a
Retzius cell is electrically stimulated with increasing frequencies,
the mucus production of the body wall area, innervated by this cell,
increases correspondingly. This shows that the integumental mucus production (probably involved in the homeostatic osmoregulatory function
of the integument) is under direct nervous control (Lent, 1973).

An involvement of nervous structures controlling the nephridial function of the leech is very probable, too, since a nerve cell has been
found (Fischer, 1960; Zerbst-Boroffka et al., 1982; Wenning, 1983)
lying on the urinary bladder sending processes into the nephridium
and into the CNS. Our ultrastructural studies demonstrated axon sections containing neurosecretory material and invading the nephridial
cells. Electrophysiological studies will elucidate the special function of the nephridial nerve cell (Wenning, 1983).

Furthermore, we could demonstrate an antidiuretic factor circulating
in the blood of leeches adapted to higher salinities (Zerbst-Boroffka
et al., 1983).

Most of the recent work on osmoregulatory function of the integument
and nephridia was done with fresh water annelids. Perhaps some of the
control mechanisms demonstrated in these annelids may function in a
similar way in estuarine annelids during physiological adaptation
to different salinities.

References

Ahearn GA and Gomme J (1975) Transport of exogenous D-glucose by the
integument of a polychaete worm (Nereis diversicolor Müller).
J Exp Biol 62: 243-264

Boroffka I (1968) Osmo- und Volumenregulation bei Hirudo medicinalis.
Z vergl Physiol 57: 348-375

Brandenburg J (1975) The morphology of the protonephridia. Fortschr
Zool 23: 1-17

Carley WW (1978) Neurosecretory control of integumental water exchange in the earthworm Lumbricus terrestris L. Gen comp Endocrin
35: 46-51

Carley WW Caracciolo EA Mason RT (1983) Cell and coelomic fluid
volume regulation in the earthworm Lumbricus terrestris. Comp
Biochem Physiol 74A: 569-575

Cornell JC (1982) Sodium and chloride transport in the isolated intestine of the earthworm, Lumbricus terrestris (L). J Exp Biol
97: 197-216

Doneen BA and Clark ME (1974) Sodium fluxes in isolated body walls of the euryhaline polychaete, Nereis (Neanthes) succinea.Comp Biochem Physiol 48A: 221-228

Ferraris JD and Schmidt-Nielsen B (1982) Volume regulation in an intertidal oligochaete, Clitellio arenarius (Müller). I Short- term effects and the influence of the supra- and subesophageal ganglia. J Exp Zool 222: 113-128

Fischer E (1969) Morphological background of the regulation of nephridial activity in the horse leech (Haemopis sanguisuga L.) Acta biol Acad Sci hung 20: 381-387

Freeman RF and Shuttleworth TJ (1977a) Distribution of dry matter between the tissues and coelom in Arenicola marina (L.) equilibrated to diluted sea water. J mar biol Ass UK 57: 97-107

Freeman RF and Shuttleworth TJ (1977b) Distribution of water in Arenicola marina (L.) equilibrated to diluted sea water. J mar biol Ass UK 57: 501-519

Freeman RF and Shuttleworth TJ (1977c) Distribution of intracellular solutes in Arenicola marina (polychaeta) equilibrated to diluted sea water. J mar biol Ass UK 57: 889-905

Gauer OH (1978) Role of cardiac mechanoreceptors in the control of plasma volume. In: Jorgensen and Skadhauge (ed.) Osmotic and volume regulation. Proceed Alfred Benzon Symp XI, Munksgaard, Copenhagen, p 229-247

Hausmann K (1982) Elektronenmikroskopische Untersuchungen an Anaitides mucosa (Annelida, Polychaeta). Cuticula und Cilien, Schleimzellen und Schleimextrusion. Helgol Meeresunt 35: 79-96

Koechlin N (1966) Ultrastructure du plexus sanguin périoesophagien, ses relations avec les néphridies de Sabella pavonina Savigny. C R Acad Sci (Paris) 262: 1266-1269

Koechlin N (1975) Micropuncture studies of urine formation in a marine invertebrate Sabella pavonina Savigny (Polychaeta, Annelida). Comp Biochem Physiol 52A: 459-464

Koechlin N (1981a) Reabsorption and accumulation of α-amino-iso-butyric acid in the nephridia of Sabella pavonina Savigny (Annelida, Polychaeta). Comp Biochem Physiol 68A: 663-667

Koechlin N (1981b) Structure and function of the nephridia in Sabella pavonina Savigny (Polychaeta,Annelida). Comp Biochem Physiol 69A: 349-355

Koechlin N (1982) Reabsorption and accumulation of glycine in the nephridia of Sabella pavonina Savigny (Annelida, Polychaeta). Comp Biochem Physiol 73A: 311-313

König ML Powell EN Kasschau MR (1981) The effect of salinity change
    on the free amino acid pools of two neireid polychaetes, Neanthes
    succinea and Leonereis culveri. Comp Biochem Physiol 70A: 631-637

Kümmel G (1975) The physiology of protonephridia. Fortschr Zool 23:
    18-32

Lent CM (1973) Retzius cells: Neuroeffectors controlling mucus re-
    lease by the leech. Science 179: 693-696

Linton LR Davies RW Wrona FJ (1982) Osmotic and respirometric res-
    ponse of two species of Hirudinoidea to changes in water chemistry.
    Comp Biochem Physiol 71A: 243-247

Machin J (1975) Osmotic responses of the bloodworm Glycera dibranchi-
    ata Ehlers: a graphical approach to the analysis of weight regula-
    tion. Comp Biochem Physiol 52A: 49-54

Malecha J (1983) Osmorégulation chez Theromyzon. Gen Comp Endocrin
    49: 344-351

Mangum CP (1982) The function of gills in several groups of inverte-
    brate animals. In: Houlihan, Rankin and Shuttleworth (ed.) Gills
    (Soc Exp Biol sem ser 16). Cambr Univ Press, London, New York,
    p 77-97

Mangum CP Dyken JA Henry RP Polites G (1978) The excretion of NH_4^+
    and its ouabain sensitivity in aquatic annelids and molluscs. J
    Exp Zool 203: 151-157

Mangum CP Saintsing DG Johnson JM (1980) The site of ion transport
    in an estuarine annelid. Mar Biol Let 1: 197-204

Nagabhushanam R and Kulkarni GK (1979) Effect of dehydration though
    dessiccation on brain neurosecretory cells of the freshwater in-
    dian leech Poecilobdella viridis (Blanchard). J Anim Morphol
    Physiol 26: 121-125

Oglesby LC (1975) An analysis of water-content regulation in selec-
    ted worms. In: Vernberg FJ (ed.) Physiological ecology of estua-
    rine organisms. Univ South Carol Press, South Carolina, p 181-204

Oglesby LC (1978) Salt and water balance. In: Mill PJ (ed.) Physiolo-
    gy of annelids Acad Press, London New York Francisco, p 555-658

Oglesby LC (1981) Volume regulation in aquatic invertebrates. J Exp
    Zool 215: 289-301

Oglesby LC Mangum CP Heacox AE Ready NE (1982) Salt and water balance
    in the polychaete Nereis virens. Comp Biochem Physiol 73A: 15-19

Peters W (1977) Possible sites of ultrafiltration in Tubifex tubifex
    Müller (Annelida Oligochaeta). Cell Tiss Res 179: 367-375

Pfannenstiel HD and Grünig C (1982) Structure of the nephridium in
    Ophryotrocha puerilis (Polychaeta,Dorvilleidae). Zoomorph 101:

187-196

Quinn RH and Bashor DP (1982) Regulation of coelomic chloride and osmolarity in Nereis virens in response to low salinities. Comp Biochem Physiol 72A: 263-265

Reynoldson TB and Davies RW (1980) A comparative study of weight regulation in Nephelopsis obscura and Erpobdella punctata (Hirudinoidea). Comp Biochem Physiol 66A: 711-714

Richards KS (1978) Epidermis and cuticle. In: Mill (ed.) Physiology of annelids. Academic Press, London New York San Francisco, p 33-61

Richards KS and Arme C (1979) Transintegumentary uptake of amino acids by the lumbricid earthworm Eisenia foetida. Comp Biochem Physiol 64A: 351-356

Richards KS and Arme C (1980) Transintegumentary uptake of D-galactose, D-fructose and 2-deoxy-D-glucose by the lumbricid earthworm Lumbricus rubellus. Comp Biochem Physiol 66A: 209-214

Schlichter LC (1982) Unstirred mucus layers: ion exchange properties and effect on ion regulation in Lymnea stagnalis. J Exp Biol 98: 363-372

Schöne C (1971) Über den Einfluß von Nahrung und Substratsalinität auf Verhalten, Fortpflanzung und Wasserhaushalt von Enchytraeus albidus Heule (Oligochaeta). Oecologia (Berl) 6: 254-266

Siebers D (1976) Absorption of neutral and basic amino acids across the body surface of two annelid species. Helgol wiss Meeresunters 28: 456-466

Siebers D and Bulnheim HP (1977) Salinity dependence, uptake kinetics and specificity of amino-acid absorption across the body surface of the oligochaete annelid Enchytraeus albidus. Helgol wiss Meeresunters 29: 473-492

Siebers D and Ehlers U (1978) Transintegumentary absorption of acidic amino acids in the oligochaete annelid Enchytraeus albidus.Comp Biochem Physiol 61A: 55-60

Skaer H Le B (1974) The water balance of a serpulid polychaete, Mercierella enigmatica (Fauvel). J Exp Biol 60: 321-370

Storch V and Alberti G (1978) Ultrastructural observations on the gills of polychaetes. Helgol wiss Meeresunters 31: 169-179

Takeuchi N (1980a) Neuroendocrine control of hydration in megascolecid earthworms. Comp Biochem Physiol 67A: 341-345

Takeuchi N (1980b) Effect of brain removal on the osmotic and ionic concentrations of the coelomic fluid of earthworms placed in soil and salt solutions. Comp Biochem Physiol 67A: 347-352

Takeuchi N (1980c) A possibility of elevation of the free amino acid
    level for the extracellular hyperosmotic adaptation of the earth-
    worm Eisenia foetida (Sav.) to the concentrated external medium.
    Comp Biochem Physiol 67A: 353-355

Takeuchi N (1980d) Control of coelomic fluid concentration and brain
    neurosecretion in the littoral earthworm Pontodrilus matsushimensis
    Iizuka. Comp Biochem Physiol 67A: 357-359

Wenning A (1983) A sensory neuron associated with the nephridia of
    the leech Hirudo medicinalis L. J comp Physiol 152: 455-458

Wenning A Zerbst-Boroffka I Bazin B (1980) Water and salt excretion
    in the leech (Hirudo medicinalis L.). J comp Physiol 139: 97-102

Wilson RA and Webster LA (1974) Protonephridia. Biol Rev 49: 127-160

Zerbst-Boroffka I (1975) Function and ultrastructure of the nephri-
    dium in Hirudo medicinalis L. III. Mechanisms of the formation of
    primary and final urine. J comp Physiol 100: 307-315

Zerbst-Boroffka I (1977) The function of nephridia in annelids. In:
    Gupta, Moreton, Oschman, Wall (ed.) Transport of ions and water in
    animals. Academic press, London New York San Francisco, p763-770

Zerbst-Boroffka I (1978) Blood volume as a controlling factor for
    body water homeostasis in Hirudo medicinalis. J comp Physiol 127:
    343-347

Zerbst-Boroffka I and Haupt J (1975) Morphology and function of the
    metanephridia in annelids. Fortschr Zool 23: 33-47

Zerbst-Boroffka I Bazin B Wenning A (1982) Nervenversorgung des Ex-
    kretionssystems und der Lateralgefäße von Hirudo medicinalis L.
    Verh Dtsch Zool Ges 1982: 341

Zerbst-Boroffka I Wenning A Bazin B (1982) Primary urine formation
    during diuresis in the leech, Hirudo medicinalis L. J comp Physiol
    146: 75-79

Zerbst-Boroffka I and Wiederholt M (1982) The mechanisms of primary
    urine formation in the nephridia of the leech during diuresis.
    IV. Europ Colloq Renal Physiol (Prag)

Zerbst-Boroffka I Wenning A Kollmann R Hildebrandt JP (1983) Anti-
    diuretischer Faktor beim Blutegel, Hirudo medicinalis. Verh Dtsch
    Zool Ges 1983

Control of the extracellular fluid osmolality in Crustaceans

A.J.R.PEQUEUX and R.GILLES

I. INTRODUCTION

Euryhaline crustaceans withstanding changes in environmental salinity exhibit almost all of the possible patterns of blood osmotic regulation (for review, see Gilles and Péqueux, 1983; Mantel and Farmer, 1983). Nevertheless, most of marine crustaceans, which means the greatest number, are osmoconformers.
A review centered on the problems related to the control of the extracellular fluids osmolality in marine euryhaline crustaceans will thus be concerned with a rather limited number of groups. It is however worth noticing that it will concern the species which have achieved, more or less sucessfully, the conquest of the widest variety of habitats. These species are able to maintain their internal osmotic concentration relatively different, either higher or lower, from that of the medium, over part or all of their ecological salinity range, despite steep concentration gradients generating important diffusive forces.

Basically, the maintenance of a relative steady state balance of the body fluids composition whatever the environmental salinity envolves two main categories of mechanisms : (1) limiting mechanisms resulting in minimizing the leaks due to the diffusive forces, and (2) energy consuming compensatory mechanisms that produce a counter movement of solute equal in magnitude to diffusive loss or gain. The hyper- and hypoosmotic states observed result from charge and discharge phenomena of inorganic ions, controlled by passive and active mechanisms at work in specialized boundary epithelia.

Early experiments performed on whole animals emphasized the part played by both categories of mechanisms but did not allow identification of organs and tissues involved. These works will not be considered here since they have been more extensively discussed in various reviews from the early sixties up to now (Gilles, 1975; Lockwood, 1977; Mantel and Farmer, 1983; Potts and Parry, 1964; Schoffeniels and Gilles, 1970 ; see also the research articles of Bryan 1960,a,b; Croghan and Lockwood, 1968; Shaw, 1961,a,b; Sutcliffe, 1968; Zanders, 1981).

A priori, several boundary epithelia could be the side of important ions movements related to blood osmoregulation. These are the body wall, the gut, the excretory organs and the gills. Up to now, informations on the part played by the body wall, the excretory organs and the gut are still very scanty. There is almost no information on the body wall, and information on the gut are far from being clear

(Ahearn 1978, 1980; Ahearn and Tornquist, 1977; Ahearn *et al*. 1977; Croghan, 1958; Dall, 1967; Farmer, 1980; Green *et al*., 1959; Mykles, 1981; Towle, 1981). If it is actually well established that the renal organ functions in volume regulation and in compensatory NaCl reabsorption in several FW hyperregulators, its contribution to total ionic balance in most estuarine hyperregulators is minimal if any (see for instance Binns, 1969a,b; Cameron and Batterton, 1978; Holliday, 1978, 1980; Kamemoto and Tullis, 1972; Riegel and Cook, 1975; Towle, 1981). In fact, gills appear to be the tissue effecting the active compensatory intake of NaCl. This review will thus be essentially concerned with some characteristics of this tissue as related to the osmoregulation process in marine euryhaline crabs.

II. THE GILL AS AN OSMOREGULATORY ORGAN

A. Structural aspects

Since the early experiments of Koch (1934), silver staining has been used to indicate regions of arthropod cuticle that are permeable to Cl⁻ and thus likely to be involved in salts movements. This technique has been used on various species of crustacea and led to identification of the gill epithelium as the primary site of blood ionic regulation (Barra *et al*., 1983; Copeland, 1968; Koch, 1934). In the strong euryhaline hyperregulator chinese crab *Eriocheir sinensis* for instance, it has been shown that the three posteriorly located pairs of gills only exhibited large silver staining patches while anterior gills had no staining regions (Barra *et al*., 1983). Rather similar observations were done also by Copeland (1968) on the landcrab *Gecarcinus lateralis*. Such results might thus be taken as an indication of structural differences between anterior and posterior gills, possibly related to functional differences.

Recently, these differences have been studied in our laboratory by electron microscopy in the case of the chinese crab *Eriocheir sinensis*. As shown in figures 1 and 2, there exists important differences in the ultrastructure of the three anterior and the three posterior pairs of gills in this species. The epithelium of the posterior gills lamellae (P.G.) is indeed much thicker (up to 10μm and more) than in anterior gills (A.G.) (only 2-4 μm), and its cuticle tends to be thinner (0.3 μm in P.G. against 1 μm in A.G.). In anterior gills, nuclei of the thin epithelial cells generally protrude within the haemolymph space. The cells' apical side just beneath the cuticle is very lightly folded which makes the extracellular compartment under the cuticle extremely reduced. The amount of intracellular organelles is very limited and the plasma membrane does not exhibit extensive latero-basal infoldings. It seems quite reasonable to consider that the main and essential functions of that tissue must be respiration only.

In crabs acclimated to dilute media, the posterior gills are essentially characterized by a complex and well-developed network of large apical evaginated and digitated folds. These membrane folds produce a large and very characteristic extracellular compartment under the cuticle. On the other hand, there are deep latero-basal folds coming into close contact with the membrane of mitochondria. Mitochondria may become very abundant, eventually completely filling the cytoplasmic space within and outside of these folds.

Although details may vary from one animal to another, a structure similar to that described in the posterior gills epithelium has been demonstrated in gills of many other osmoregulator crustaceans and in most of the organs responsible for hydromineral regulation in other animal groups (Bielawski, 1971; Bulger, 1963; Copeland, 1964, 1968;

Copeland and Fitzjarrell, 1968; Foster and Howse, 1978; Komnick, 1963;
Philpott and Copeland, 1963; Tandler, 1963). It is characteristic of
so-called "salt-transporting epithelia" (for review see Gilles and
Péqueux, 1981; Berridge and Oschman, 1972).

Evidences support the idea that changes in the ultrastructure of
that epithelium, possibly correlated with changes in its physiological
function, occur upon acclimation to media of different salinity.
Copeland and Fitzjarrell (1968) indeed demonstrated that silver-stain-

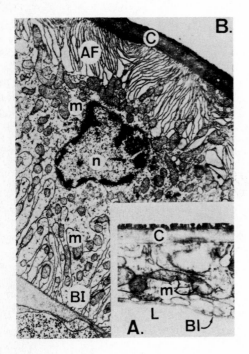

Fig.1 : *Ultrastructure of epithelial
cells from gill's lamellae of
Eriocheir sinensis acclimated to
freshwater (X 23,150)*
A : *anterior gill*
*Cuticular surface at top (c); thick
basal lamina (B.l) lining the haemolymph lacuna (L) below; mitochondria
(m).*
B : *posterior gill*
*Apical membranes extensive folds(AF);
basal and latero-basal membranes
interdigitations (BI) coming into
close contact with mitochondria (m);
cuticle (c); nucleus (n).*
*(After Péqueux and Barra, unpublis-
hed; see acknowledgments).*

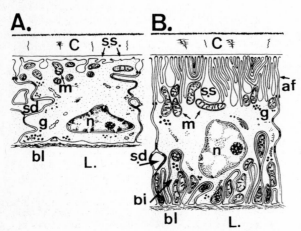

Fig.2 : *Diagrammatic
drawings of cross sections
of anterior (A) and posterior (B) gills lamellae
of Eriocheir sinensis acclimated to freshwater.
Apical folds (a.f); basal
infoldings(b.i); basal
lamina(b.l); cuticle (c);
glycogene(g);haemolymph (L);
mitochondria(m); nucleus(n);
septate desmosomes(s.d);
sub-cuticular space (s.s.)
(After Barra , unpublished;
see acknowledgments).*

ing patches of the gills of hyper-osmotic regulators increased in size after acclimation to dilute media. In the chinese crab *E.sinensis*, acclimation to SW results in a severe decrease of the area of the cells apical surface, mainly due to the shortening of apical invaginations and disappearance of the apical cellular labyrinth (figure 3). Decrease in width of the intercellular spaces and changes in the structure of mitochondria have also been quoted in the review by Mantel and Farmer, 1983.

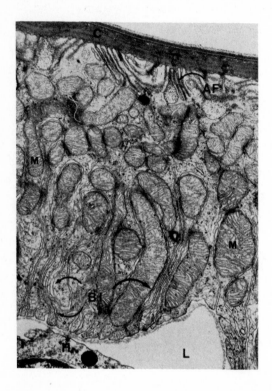

Fig. 3 : Ultrastructure of posterior gill epithelial cells from Eriocheir sinensis acclimated to sea water (X 19,000). Explanations : see the text and the legend of Figures 1 and 2. (After Péqueux and Barra, unpublished; see acknowledgments)

If one considers the thick epithelium of the posterior gills to be the only one implicated in transepithelial salt active uptake processes, this should indicate that ions active transport mechanisms involved in the control of blood osmolality in *E.sinensis* must be restricted essentially to the posterior pairs of gills. That this is indeed the case has been demonstrated by flux studies on a perfused preparation of isolated gills (Péqueux and Gilles, 1978, 1981; Gilles and Péqueux, 1981). Let us now consider the results of these experiments.

B. Physiological aspects

From considerations based on the comparison between the ratios that can be calculated from the actual flux measurements and those calculated according to the Ussing's equation for passive movements of an ion, it has been established that Na^+ fluxes in the Anterior gills of the euryhaline chinese crab *E.sinensis* are essentially passive (Péqueux and Gilles, 1981). Moreover , the study of the fluxes magnitude as a function of the external Na^+ concentration reveals a saturation kinetics which indicates that the movements of Na^+ across the anterior gills epithelium are not due to a simple diffusional process but rather

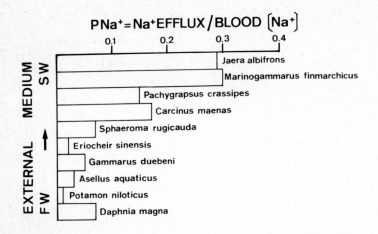

Fig.4 : Sodium permeability in some crustaceans
acclimated to different salinities.
(drawn according to data from Harris, 1972;
Shaw, 1961a,b; Sutcliffe, 1968 and Rudy, 1966).

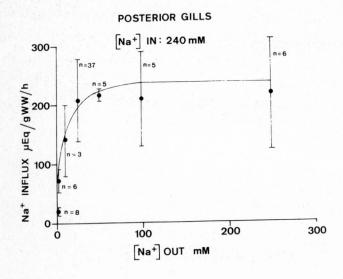

Fig.5 : Relation between external Na$^+$ concentration
(abscissa) and Na$^+$ influx (ordinate) in perfused
posterior gills isolated from FW-acclimated chinese
crabs Eriocheir sinensis. Mean values of n experiments ± standard deviation (S.D.).
(After Péqueux and Gilles, 1981).

are carrier-mediated (Péqueux and Gilles, 1981). Further, the efflux of Na^+ largely decreases when the external Na^+ concentration decreases, in spite of the tremendous increase of the transepithelial Na^+ gradient. This supports the idea that the Na^+ permeability of Anterior gills decreases at low external Na^+ level, a situation which is of interest since it should decrease the salt loss occurring in animals acclimating to low salinities (Péqueux and Gilles, 1981). Such a situation moreover comes within the scope of the limiting mechanisms involved in minimizing the leaks due to the diffusive forces and corroborates the observations done on whole animals (Figure 4).

In Posterior gills , the curve showing the dependency of the Na^+ influx upon the Na^+ concentration of the external medium also reveals the existence of a carrier-mediated mechanism, but, whatever the amplitude of the Na^+ gradient across the epithelium, it never has been possible to demonstrate any significant efflux of Na^+ (Péqueux and Gilles, 1981) (Figure 5). When considered in the context of the Ussing's equation, the Na^+ entry across the Posterior gills epithelium must therefore be active.

Up to now, the nature of the system(s) implicated in the gills' active transport process(es) has remained much disputed and, in many instances, is essentially matter of speculation. Early evidences, based on "*in vivo*" studies, suggest the existence of independent mechanisms for the absorption of Na^+ and Cl^- in dilute media (Krogh, 1939; Shaw, 1960, 1961), both mechanisms having different rate constants and electrical neutrality thus requiring counterions of same sign. In the crayfish, results on ammonia production in relation to the salinity of the acclimation medium have been interpreted as indicating a correlation between Na^+ influx and NH_4^+ movements in the opposite direction and the concentration in the environmental medium, hence the early idea of a direct coupling between Na^+ and NH_4^+ fluxes (Shaw, 1960). That idea has been furthermore substantiated by the important increase of NH_4^+ blood level and ammonia output which occurs during acclimation to dilute media (Gérard and Gilles, 1972; Mangum *et al.*, 1976; see Gilles, 1975 for review). Results of our experiments on isolated perfused gills of *Eriocheir sinensis* are in agreement with the idea that some coupling between Na^+ and NH_4^+ movements might occur in the posterior gills (Gilles and Péqueux, 1981; Péqueux and Gilles, 1978b, 1981). The NH_4^+ efflux is indeed ouabain sensitive; it shows some dependency on the external concentration of Na^+ and substitutes quite well for K^+ in the assay of $(Na^+ + K^+)$ATPase activity of membrane fractions. However, there is such a large discrepancy between the flux magnitude of both ions and their dependency on external Na^+ that the hypothesis of an only and necessary coupling is unlikely (figure 6). Moreover, increasing the concentration of NH_4Cl in the perfusate always failed to induce any significant modification of the Na^+ influx (Péqueux and Gilles, 1981).

The major part of the Na^+ influx must therefore be accounted for by another process which could be an electrogenic system or another kind of exchange. Up to now, this still remains matters of speculation. Some evidences suggest however that H^+ ions may be considered as likely potential candidates for that exchange but more results are needed to assess this hypothesis (Ehrenfeld 1974; Kirschner *et al.* 1973; Shaw, 1960).

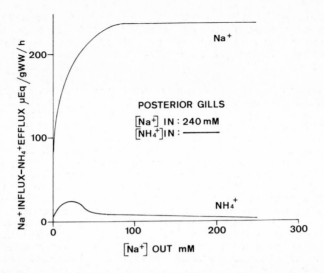

Fig.6 : Relation between external Na⁺ concentration
(abscissa), Na⁺ influx and NH₄⁺ efflux (ordinate) in
perfused posterior gills isolated from FW-acclimated
chinese crabs E.sinensis. (After Péqueux and Gilles,
1981).

C. Biochemical Aspects

Up to now, little work has been devoted to the biochemical
problems related to the transport activity and to osmoregulation in
crustaceans. These problems are of three types : 1) How does correlate
the gills energy metabolism and the transport activity; 2) What are
the molecular supports of the transport activities; 3) What are the
effects of changes in blood ions concentrations on proteic structures
implicated in metabolic processes both extra and intracellular; this
last question arises from the fact that the blood NaCl content of most
euryhaline crustaceans undergoes large fluctuations when the animals
are acclimated to media of different salinities.

With an exception of a few studies on the effects of NaCl on some
enzymes activity (Gilles, 1974a,b), not much has been done in relation
with the effect of NaCl on metabolic structures in crustaceans. We will
thus concentrate on some problems related to the energy metabolism.
Early experiments tryed to correlate acclimation to different media and
the concomitant requirements for ionoregulation with the oxydative
metabolism and the oxygen consumption. In most cases, acclimation to
reduced salinity induces an increase in oxygen consumption as well as
in oxydative metabolism. This has been shown repeatedly on whole animals as well as on isolated tissues (Engel et al., 1974, 1975;
Florkin, 1960; King, 1965; Kinne, 1971; Vernberg, 1956; for review see
Gilles 1975; Mantel and Farmer, 1983). The fundamental significance of
these results have been and is still very disputed. It is however reasonable to consider at first glance that such increase in metabolic rates
should be, partly to an increase in the energy demand due to the enhancement of the activity of the transport processes. In connexion with this,
the amount of ATP as well as its turnover rate appeared to be salinity
dependent in isolated gills of the blue crab Callinectes

sapidus (Engel *et al.*, 1974, 1975). We therefore undertook a more detailed study of the pattern of purine nucleotides as related to the salinity of the acclimation medium in individual gills of two euryhaline crab species : *Eriocheir sinensis* and *Carcinus maenas* (Wanson, 1983; Wanson *et al.*, 1983).

In both crabs, the total amount of purine nucleotides reaches an order of magnitude of 6 µM/mg DNA, exhibiting great differences between the various nucleotides. As an example, in posterior gills of SW acclimated *C.maenas* : GTP = 0.95 , GDP = 0.05, IMP = 5.41, ATP = 3.75, ADP = 0.48 , AMP = 0.11 µM/mg DNA.
Whatever the acclimation salinity, a significantly higher level of adenylates has always been measured in the three posterior pairs of gills. In *C.maenas*, acclimation to dilute media results in a decrease in the total amount of adenylate nucleotides. In *E.sinensis* in turn, the adenylates amount tends to increase (Table 1). On the other hand, acclimation to dilute media results, in both species, in a significant drop, in the three posterior pairs of gills only, of the energy charge calculated according to Atkinson (A.E.C) (Table 2) (Wanson, 1983; Wanson *et al.*, 1983). This decrease seems to be mainly due to an important raise of the AMP content. The adenylate energy charge AEC= (ATP + 1/2ADP/ATP + ADP + AMP) is widely considered as a relevant measure of the metabolic energy pool available to the cell from the adenylate system (Atkinson, 1977). In that view, AEC data suggest that ATP utilization significantly overcomes ATP production in posterior gills of crabs facing a dilution stress. That conclusion is in quite good agreement with the finding that only the posterior gills of FW acclimated *E.sinensis* can actively take up Na^+. The fact that the observed drop in AEC is larger in *E.sinensis* than in *C.maenas* may be related to their different ecophysiological possibilities and to the media to which they have been acclimated. The chinese crab is indeed a very strong hyperosmoregulator and has been adapted to fresh water for these experiments while the shore crab, a much weaker regulator, has been adapted only to 1/3 sea water. The dilution stress is therefore larger in *E.sinensis*. These results thus demonstrate that the increase in energy demand that was supposed to occur in dilute media can be traced down to the level of the adenylates. The observed changes in the gills'energy charge are in good agreement with the idea that acclimation to dilute media of hyperosmoregulating crabs leads to an increase in pumping activity that can be related to a decrease in adenylates energy charge; this decrease in turn induces, the well known increase in oxydative metabolism and in oxygen consumption.

Another interesting finding that emerged from our study of the purine nucleotides patterns lies in the amount of IMP found in both species. The IMP level is indeed much larger in the gills of *Carcinus maenas* than in those of *Eriocheir sinensis*. This probably indicates differences in the pathways of the adenylates catabolism in both species. Actually not much is knownabout the purines metabolism in crustaceans and about its possible relation with the deamination mechanisms. Further studies are needed to bring some light on these problems. It is however worth noticing to consider in this context that in *C.maenas*, the level of IMP, which is the deamination product of AMP and a key metabolite in the deaminating purines cycle, is very sensitive to salinity (figure 7).

Let us now consider the possible relations between salinity-induced changes in Na^+ pumping activity and (Na^++K^+)ATPase activity. The significance of the (Na^++K^+)ATPase activity of plasma membranes from transporting tissues remains still actually much questioned; several evidences however support the idea of its direct linkage with Na^+ active transport processes. Since controversial aspects of

TABLE 1 : Effect of salinity changes on total amount of adenylates
ΣA^{*} = ATP + ADP + AMP in Anterior (AG) and Posterior (PG)
gills of the euryhaline decapods *C.maenas* and *E.sinensis*.

	CARCINUS MAENAS		ΣA^{*} (µM/MgDNA)	ERIOCHEIR SINENSIS		
	Anterior gills	Posterior gills		Anterior gills	Posterior gills	
SW	2.252 ± 0.647 N=8	3.919 ± 1.125 N=9		1.162 ± 0.468 N=8	2.546 ± 0.909 N=9	SW
SW/3	1.777 ± 0.642 N=8	2.315 ± 0.259 N=6		1.733 ± 0.988 N=7	3.944 ± 1.501 N=9	FW

Mean values of n experiments ± S.D. (courtesy of Wanson; Wanson, 1983;
Wanson *et al.*, 1983; see also acknowledgments).

TABLE 2 : Effect of salinity on the adenylate energy charge (A.E.C)
of anterior (AG) and posterior (PG) gills of *C.maenas* and *E.sinensis*.

	CARCINUS MAENAS		A.E.C.	ERIOCHEIR SINENSIS		
	Anterior gills	Posterior gills		Anterior gills	Posterior gills	
SW	0.873 ± 0.092 N=9	0.911 ± 0.031 N=9		0.874 ± 0.042 N=9	0.929 ± 0.024 N=9	SW
SW/3	0.907 ± 0.023 N=6	0.832 ± 0.031 N=6		0.917 ± 0.039 N=9	0.751 ± 0.069 N=9	FW
T-TEST	−	*		−	*	

Mean values ± S.D. of n experiments. *Significant difference at the 1%
level. A.E.C. = ATP + 1/2ADP/ ATP + ADP + AMP.
(After Wanson, 1983; Wanson *et al.*,1983; see also acknowledgments).

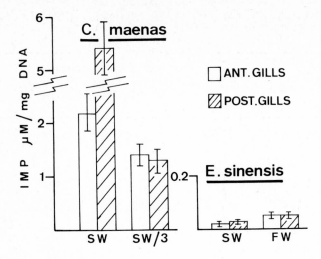

Fig.7 : IMP content of anterior and posterior
gills of C.maenas and E.sinensis at different
salinities of the environment.
Mean values ± S.D. of at least 6 experiments.
(After Wanson, 1983)

$(Na^{+}+K^{+})$ATPase function in marine and estuarine animals will be considered more extensively in another part of this volume (see Towle,
chapter), we will examine that question only briefly and restrict
our analysis to the regulating species described above and studied
in our laboratory.

To begin with, let us recall that the existence of a $(Na^{+}+K^{+})$ATP
ase activity in crab gills has been investigated by several authors
and is actually rather well documented (**Mantel** and Olson, 1976;
Neufeld *et al.*, 1980; Péqueux and Gilles, 1977; Péqueux and Chapelle,
1982; Péqueux *et al.*, 1983; Péqueux *et al.*, 1984; Towle *et al.*,1976;
Spencer *et al.*, 1979). In both hyperosmoregulators *E.sinensis* and
C.maenas like in several other species, a high $(Na^{+}+K^{+})$ATPase activity
characterizes the gill epithelium and the level of activity has been
shown to be dependent on the gill type considered (Péqueux *et al.*,1984;
Wanson , Péqueux and Gilles, unpublished).The three most posterior
pairs of gills indeed exhibit the highest activity (Figure 8). It
seems reasonable to consider that these differences reflect functional
differences related to the localization of the transepithelial active
transport of Na^{+}.

As shown in Figure 9, acclimation to media of reduced salinity
results in a significant increase of the enzyme activity.
No significant differences has been detected between the kinetics characteristics of the enzyme from SW or FW acclimated *E.sinensis*
(Péqueux *et al.*,1984). Up to now, the question of the origin of salinity induced changes of activity is not yet solved. In the chinese crab
at least, it can be considered that acclimation to media of different
salinity does not induce synthesis of kinetically different $(Na^{+}+K^{+})$

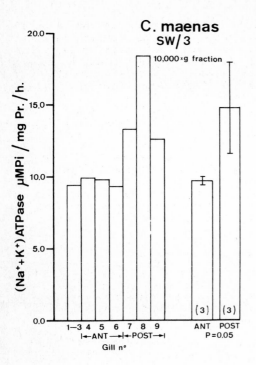

Fig.8 : $(Na^+ + K^+)ATPase$ activity
of individual gills (typical
results) and in so-called anterior (ANT) and posterior (POST)
gills (mean data ± SD) of the
shore crab Carcinus maenas acclimated to dilute sea water (SW/3).
(Courtesy of Wanson, unpublished
results).

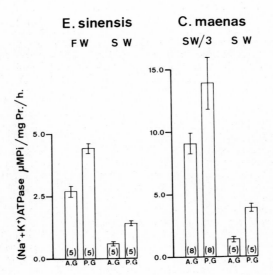

Fig.9 : $(Na^+ + K^+)ATPase$ activity in the 10,000Xg
fraction of anterior (AG) and posterior (PG)
gills of the euryhaline crabs C.maenas and E.sinensis
acclimated to concentrated (SW) and dilute (SW/3, FW)
media (After Wanson, Péqueux and Gilles, unpublished
results).

ATPase but rather modifies the amount of enzyme molecules and/or activates preexisting ones. Another interesting point to consider is that the most important changes in the enzyme specific activity occur in the 10,000g fraction which is rich in mitochondria and membrane fragments. This could suggest that the enzyme activity related to the blood hyperosmotic regulation in *E.sinensis* could be essentially restricted to the 10,000g fraction of the posterior gills.

In that view, the (Na^++K^+) ATPase activity of the 100,000g fraction as well as the activity in anterior gills would not be primarily associated with the transepithelial movement of ions but rather with the control of the intracellular levels of Na^+ and Cl^-.

This structural organization could possibly be the support, in the posterior gills of *E.sinensis*, of a specific pathway and Na^+ pool implicated in the transepithelial movements of that ion. This pathway would be independent of the main, intracellular Na^+ pool. That idea is still up to now highly speculative, but experiments are actually in progress in our laboratory to evaluate that hypothesis.

III. CONCLUSIONS

Phenomenologically, the effects of salinity changes on the hemolymph osmolality and ions content in Crustaceans may be actually considered as very well documented. However, a lot of questions remain to be answered in every area, particularly concerning the location and precize nature of control mechanisms.

From the biochemical point of view, the molecular basis of the processes involved still remain quite unclear, and almost nothing is known concerning a possible hormonal action and the way the regulation of ions movements is achieved.

Moreover, a distinction should be done between short-term and long-term acclimation to fluctuating salinities.

Most of the current understanding of the mechanisms responsible for the anisosmotic regulation of the body fluids osmolality in crustaceans is based on studies of a restricted number of decapods or on some very specialized forms. Moreover, most of these studies suffer the disadvantage to be performed on whole animals and do not allow any discrimination between the various processes involved.

From this review, it appears that, in the recent years, a great deal of informations in that field arised from experimentation with *in vitro* preparation.

These experiments emphasized the prominent part played by the gills in the control of the body fluids composition, but also established the existence of some kind of specialization in that function. As a general rule, the gills located posteriorly in the gill chamber of euryhaline hyperregulators appeared to be the main if not the only ones responsible for active Na^+ uptake in dilute media. Evidences support only partly the working hypothesis that ammonia can be used as a counterion in the active uptake of Na^+ at the gill level. However, that participation in the total influx of Na^+ has been proved to remain quite small in the chinese crab *E.sinensis*. Whether the Na^+ uptake is due to another kind of coupled transport or to an electrogenic system are thus questions that are still matters of speculation.

Good correlations have been established, in euryhaline species like the chinese crab *E.sinensis*, between the ultrastructure of the gill epithelium as revealed by electronmicroscopy, its transport properties and its biochemical characteristics.

Extended to other species of Crustaceans, further investigations with *in vitro* preparation, but also concerted study in biochemistry, physiology, morphology and ecology should lead to identification of the mechanisms which are responsable either of the complete or relative success, either of the check of species conquesting media of fluctuating salinities.

ACKNOWLEDGEMENT

Part of the work described in this paper has been aided by grants "Crédits aux Chercheurs" from the Fonds National de la Recherche Scientifique n° 15.422.82 F to R.G.

We wish to thank Drs R.Kirsch and J.A.Barra from the University of Strasbourg (France), who did the electronmicroscopy study in the course of a collaboration with us.

Many thanks also to Dr.C.Leray who welcomed us in his CNRS laboratory of Strasbourg for determinating nucleotides patterns in crab gills by HPLC and to Mrs. C.Gutbier for efficient technical assistance.

REFERENCES

Ahearn G.A. (1978).Allosteric co-transport of sodium, chloride and calcium by the intestine of freshwater prawns. J. Membrane Biol. 42 : 281-300.

Ahearn G.A. (1980).Intestinal electrophysiology and transmural ion transport in freshwater prawns. Am. J. Physiol. 239 : C1-10.

Ahearn G.A., Tornquist A. (1977). Allosteric cooperativity during intestinal cotransport of sodium and chloride in freshwater prawns. Biochem. Biophys. Acta 471 : 273-279.

Ahearn G.A., Maginniss L.A., Song Y.K., Tornquist A. (1977). Intestinal water and ion transport in freshwater malacostracan prawns (Crustacea). In "Water relations in Membrane transport in Plants and Animals" (Jungreis A.,Hidges T., Kleinzeller A. and Schultz S., eds.) Academic Press, New York. p. 129-142.

Atkinson D.E. (1977). Cellular energy metabolism and its regulation. Academic Press. New York.

Barra J.A., Péqueux A., Humbert W. (1983). A morphological study on gills of a crab acclimated to fresh water. Tissue and cell 15(4) : 583-596.

Berridge M.J., Oschman J.L. (1972). Transporting epithelia. Academic Press, New York and London.

Bielawski J. (1971). Ultrastructure and ion transport in gill epithelium of the crayfish, *Astacus leptodactylus* Esch. Protoplasma 73 : 177-190.

Binns R. (1969a). The physiology of the antennal gland of *Carcinus maenas* (L.). I. The mechanism of urine production. J. Exp. Biol. 51 : 1-10.

Binns R. (1969b). The physiology of the antennal gland of *Carcinus maenas* (L.).II. Urine production rates. J.Exp.Biol. 51 : 11-16.

Bryan G.W. (1960a). Sodium regulation in the crayfish *Astacus fluviatilis*. II. Experiments with sodium depleted animals. J. Exp. Biol. 37 : 83-89.

Bryan G.W. (1960b). Sodium regulation in the crayfish *Astacus fluviatilis*. III. Experiments with NaCl-loaded animals. J. Exp. Biol. 37 : 113-128.

Bulger R.E. (1963). Fine structure of the rectal (salt-secreting) gland of the spiny dogfish, *Squalus acanthias*. Anat. Rec. 147 : 95-127.

Cameron J.N., Batterton C.V. (1978). Antennal gland function in the freshwater blue crab, *Callinectes sapidus* : water, electrolyte, acid-base and ammonia excretion. J. Comp. Physiol. 123 : 143-148.

Copeland D.E. (1964). Salt absorbing cells in gills of crabs *Callinectes* and *Carcinus*. Biol. Bull. 127 : 367-368.

Copeland D.E. (1968). Fine structure of salt and water uptake in the land-crab *Gecarcinus lateralis*. Am. Zool. 8 : 417-432.

Copeland D.E., Fitzjarrell A.T. (1968). The salt absorbing cells in the gills of the blue crab (*Callinectes sapidus*, Rathbun) with notes on modified mitochondria. Z. Zellforsch. Mikrosk. Anat. 92 : 1-22.

Croghan P.C. (1958). The mechanism of osmotic regulation in *Artemia salina* (L.) : The physiology of the gut. J. Exp. Biol. 35 : 243-249.

Croghan P.C. , Lockwood A.P.M. (1968). Ionic regulation of the Baltic and freshwater races of the isopod *Mesidotea (Saduria) entomon* (L.). J. Exp. Biol. 48 : 141-158.

Dall W. (1967). Hypo-osmoregulation in Crustacea. Comp. Biochem. Physiol. 21 : 653-678.

Ehrenfeld J. (1974). Aspects of ionic transport mechanisms in crayfish *Astacus leptodactylus*. J. Exp. Biol. 61 : 57-70.

Engel D.W., Eggert L.D. (1974). The effect of salinity and sex on the respiration rates of excised gills of the blue crab *Callinectes sapidus*. Comp. Biochem. Physiol. 47A : 1005-1011.

Engel D.W., Ferguson R.L., Eggert L.D. (1975). Respiration and ATP concentrations in the excised gills of the blue crab as a function of salinity. Comp. Biochem. Physiol. 52A : 669-673.

Foster C.A., Howse H.D. (1978). A morphological study on gills of the brown shrimp, *Penaeus aztecus*. Tissue and Cell 10 : 77-92.

Gérard J.F. , Gilles, R. (1972). The free amino-acid pool in *Callinectes sapidus* (Rathbun) tissues and its role in the osmotic intracellular regulation. J. Exp. Mar. Biol. Ecol. 10 : 125-136.

Gilles R. (1974a). Studies on the effect of NaCl on the activity of *Eriocheir sinensis* glutamate dehydrogenase. Int. J. Biochem. 5 : 623-628.

Gilles R. (1974b). Métabolisme des acides aminés et contrôle du volume cellulaire. Arch. Internat.Physiol.Bioch. 82 : 423-589.

Gilles R. (1975). Mechanisms of ion and osmoregulation. In "Marine Ecology" (O.Kinne, ed.). Vol.2, part 1. pp.259-347.

Gilles R., Péqueux A. (1981). Cell volume regulation in crustaceans : Relationship between mechanisms for controlling the osmolality of extracellular and intracellular fluids. J.Exp.Zool. 215 : 351-362.

Gilles R., Péqueux A. (1983). Interactions of chemical and osmotic regulation with the environment. In "The Biology of Crustacea" (Dorothy Bliss, editor in chief), vol.8 (Vernberg F.J. , Vernberg W.B., eds). Academic Press. p. 109-177.

Green J.W., Harsch H., Barr L., Prosser C.L.(1959). The regulation of water and salt by the fiddler crabs *Uca pugnax* and *Uca pugilator*. Biol. Bull. (Woods Hole, Mass.) 116 : 76-87.

Farmer L. (1980). Evidence for hyporegulation in the calanoid copepod *Acartia tonsa*. Comp. Biochem. Physiol. 65A : 359-362.

Florkin M. (1960). Ecology and metabolism. In "The Physiology of Crustacea" (Waterman T.H., ed.) Vol.I. Academic Press, New York. pp.395-410.

Harris R.R. (1972). Aspects of sodium regulation in a brackish-water and a marine species of the isopod genus *Sphaeroma*. Mar. Biol. 12 : 18-27.

Holliday C.W. (1978). Aspects of antennal gland function in the crab, *Cancer magister* (Dana). Ph.D. Dissertation, University of Oregon Eugene, Oregon.

Holliday C.W. (1980). Magnesium transport by the urinary bladder of the crab, *Cancer magister*. J. Exp. Biol. 85 : 187-201.

Kamemoto F.I., Tullis R.E.(1972). Hydromineral regulation in decapod crustacea. Gen. Comp. Endocrinol. 3 : 299-307.

King E.N.(1965). The oxygen consumption of intact crabs and excised gills as a function of decreased salinity. Comp. Biochem. Physiol. 15 : 93-102.

Kinne O. (1971). Salinity : Animals - Invertebrates. In "Marine Ecology" (Kinne O., Ed.) Vol.I, Environmental Factors, Part 2. Wiley, London. pp. 821-995.

Kirschner L.B., Greenwald L., Kerstetter T.H. (1973). Effect of amiloride on sodium transfer across body surfaces of fresh water animals. Am. J. Physiol. 224 : 832-837.

Koch H.J. (1934). Essai d'interprétation de la soi-disant réduction vitale des sels d'argent par certains organes d'Arthropodes. Ann.Soc.Sci.Med.Nat.Brux., Ser.B , 54 : 346-361.

Komnick H. (1963). Electroneumikroskopische Untersuchungen zur funktionelle Morphologie des Ionentransportes in der Salzdrüse von *Larus argentatus* I. Teil : Bau und Feinstruktur der Salzdrüse. Protoplasma 56 : 274-314.

Krogh A.(1939). "Osmotic Regulation in Animals". Cambridge University Press, Cambridge.

Lockwood A.P.M. (1977). Transport and osmoregulation in Crustacea. In "Transport of Ions and Water in Animals (Gupta B.L., Oschman J.L., Moreton R.B., Wall B.J., eds). Academic Press, London. pp.673-707.

Mangum C.P., Silverthorn S.U., Harris J.L., Towle D.W., Krall A.R. (1976). The relationship between blood pH, ammonia excretion and adaptation to low salinity in the blue crab *Callinectes sapidus*. J. Exp. Zool. 195 : 129-136.

Mantel L.H., Farmer L.L. (1983). Osmotic and Ionic Regulation. In "The Biology of Crustacea" (Dorothy Bliss, Editor-in-Chief), Vol.5 (Mantel L.H.,Ed.). Academic Press. pp.53-161.

Mantel L.H., Olson J.R. (1976). Studies on the Na^+-K^+activated ATPase of crab gills. Am. Zool. 16: 223.

Mykles D.L. (1981). Ionic requirements of transepithelial potential difference and net water flux in the perfused midgut of the American lobster, *Homarus americanus*. Comp. Biochem. Physiol. 69A : 317-320.

Neufeld G.J., Holludag C.W., Pritchard J.B.(1980). Salinity adaptation of gill Na, K-ATPase in the blue crab, *Callinectes sapidus*. J. Exp. Zool. 211 : 215-224.

Péqueux A., Chapelle S.(1982). $(Na^+ + K^+)$ATPase activity and phospholipids in two euryhaline crabs as related to changes in the environmental salinity. Mar. Biol. Lett. 3 : 43-52.

Péqueux A., Gilles R.(1977). Osmoregulation of the chinese crab *Eriocheir sinensis* as related to the activity of the $(Na^+ + K^+)$ ATPase. Arch. Internat. Physiol. Biochim. 85 : 426-427.

Péqueux A., Gilles R.(1978a). Osmoregulation of the euryhaline chinese crab *Eriocheir sinensis*. Ionic transports across isolated perfused gills as related to the salinity of the environment. In "Physiology and Behaviour of Marine Organisms".(Mc Lusky,D.S., Berry A.J., eds). Pergamon Press, Oxford, New York. pp.105-111.

Péqueux A., Gilles R.(1978b). Na^+/NH_4^+ co-transport in isolated perfused gills of the chinese crab *Eriocheir sinensis* acclimated to fresh water. Experientia 34 : 1593-1594.

Péqueux A., Gilles R.(1981). Na^+ fluxes across isolated perfused gills of the chinese crab *Eriocheir sinensis*. J. Exp. Biol. 92 :173-186.

Péqueux A., Marchal A., Wanson S., Gilles R.(1984). Kinetics characteristics and specific activity of gill $(Na^+ + K^+)$ATPase in the euryhaline chinese crab, *Eriocheir sinensis* during salinity acclimation. Mar. Biol. Lett., in the Press.

Péqueux A., Chapelle S., Wanson S., Goffinet G., François C.(1983). $(Na^+ K^+)$ATPase activity and phospholipid content of various fractions of the posterior gills of *Carcinus maenas* and *Eriocheir sinensis*. Mar. Biol. Lett. 4 : 267-279.

Philpott C.W. , Copeland D.E.(1963). Fine structure of the chloride cells from three species of *Fundulus*. J. Cell. Biol. 18 : 389-404.

Potts W.T.W. , Parry G.(1964). Osmotic and Ionic regulation in animals. Pergamon Press, Oxford.

Riegel J.A., Cook M.A.(1975). Recent studies of excretion in Crustacea. In "Excretion" (Wessing A., Ed.). Fortschritte der Zoologie 23. Gustav Fisher Verlag, Stuttgart, Germany. pp.48-75.

Rudy P. (1966). Sodium balance in *Pachygrapsus crassipes*. Comp. Biochem. Physiol. 18 : 881-907.

Schoffeniels E., Gilles R.(1970). Osmoregulation in aquatic arthropods. In "Chemical Zoology" (Florkin M., Scheer B.T., Eds). Vol.V, Part A. Academic Press, New York. pp.255-286.

Shaw J. (1960). The absorption of ions by the crayfish. II. The effect of external anion. J.Exp.Biol. 37 : 534-547.

Shaw J. (1961a). Studies on ionic regulation in *Carcinus maenas*. J. Exp. Biol. 38 : 135-152.

Shaw J. (1961b). Sodium balance in *Eriocheir sinensis*. (M.Edw.). The adaptation of the Crustacea to fresh water. J. Exp. Biol. 38 : 153-162.

Spencer A.Mc.D., Fielding A.H., Kamemoto F.I. (1979). The relationship between gill Na, K-ATPase activity and osmoregulation capacity in various crabs. Physiol. Zool. 52: 1-10.

Sutcliffe D.W. (1968). Sodium regulation and adaptation to fresh water in gammarid crustaceans. J. Exp. Biol. 48 : 359-380.

Tandler B.(1963). Ultrastructure of the human submaxillary gland.
    II. The base of the striated duct cells. J. Ultrastruct. Res.
    9 : 65-75.

Towle D.W. (1981). Transport related ATPases as probes of tissue
    function in three terrestrial crabs of Palan.
    J. Exp. Zool. 218 : 89-95.

Towle D.W., Palmer G.E., Harris J.L.(1976). Role of gill Na^++K^+-
    dependent ATPase in acclimation of blue crabs (*Callinectes
    sapidus*) to low salinity. J. Exp. Zool. 196 : 315-322.

Vernberg F.J. (1956). Study of the oxygen consumption of excised
    tissues of certain marine decapod crustacea in relation to
    habitat. Physiol. Zoöl. 29 : 227-234.

Wanson S.(1983). Purine nucleotides patterns and energy charge in
    Carcinus maenas and *Eriocheir sinensis* gills during osmoregu-
    lation. In "Physiological and Biochemical Aspects of Marine
    Biology", Abstracts volume of the 5th Conference of the ESCPB
    Taormina, Italy. pp.154-155.

Wanson S., Péqueux A., Leray, C.(1983). Effect of salinity changes
    on adenylate energy charge in gills of two euryhaline crabs.
    Arch. Internat. Physiol. Bioch. 91(2) : B81-B82.

Zanders I.P. (1981). Control and dynamics of ionic balance in *Carcinus
    maenas*(L.). Comp. Biochem. Physiol. 70A : 457-468.

Osmotic and ionic regulation in saline-water Mosquito larvae

T.J.BRADLEY, K.STRANGE and J.E.PHILLIPS

I. Regulatory Capacities and Sites of Exchange

The larvae of a number of species of mosquitoes in the genus *Aedes* can survive and complete development in waters ranging in osmotic concentration from distilled water to several times the concentration of seawater. This amazing range of tolerance is achieved by strict hyper-regulation in dilute waters and hyporegulation in concentrated media, such that a fairly constant hemolyph ionic concentration is maintained (Fig. 1a). Larvae of the species *Aedes taeniorhynchus* are able to survive in seawater which has been concentrated to three times its normal value (Nayar & Sauerman, 1974). Several other species are able to hyporegulate in media with an osmotic concentration two to four times that of the hemolymph (*Aedes campestris*, Kiceniuk & Phillips, 1974, Bradley & Phillips, 1977a; *A. detritus*, Beadle, 1939; Ramsay, 1950; *A. dorsalis* Sheplay & Bradley, 1982; Strange et al, 1982; *Aedes togoi*, Asakura, 1980; *Opifex fuscus*, Nicholson, 1972.) Perhaps even more remarkably, a number of species of saline-water mosquito larvae can survive in inland hypersaline ponds where ionic ratios differ substantially from seawater, e.g. ponds rich in $NaHCO_3$ or $MgSO_4$ and Na_2SO_4 (Fig. 1b) (Kiceniuk & Phillips, 1974; Bradley & Phillips, 1977a; Strange et al, 1982).

Hyporegulation by saline-water mosquito larvae is not dependent on a highly impermeable integument. The osmotic permeablity of the external cuticle has been shown to be 4.8 X 10-3 and 2 X 10-3 cm/h in *Opifex fuscus* and *A. campestris*, respectively (Nicholson & Leader, 1974; Phillips et al, 1978). These values are only 3-10 times lower than values obtained for larvae of freshwater species. As a result, larvae in saline media undergo constant osmotic loss of water from the hemolymph to the external environment. This water is replaced by

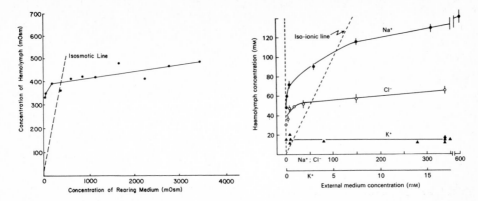

Fig. 1. (a) The relation between osmotic concentrations in the rearing media and the larval hemolymph of *A. taeniorhynchus*. Redrawn from Nayar and Sauerman (1974). (b) Ionic concentrations in hemolymph of *A. campestris* larvae as a function of external levels (Phillips et al., 1978).

drinking. Larvae ingest the external medium at a rate ranging from 130% of body volume per day in *Aedes dorsalis* to 240% in *A. taeniorhynchus*. These rates of drinking far exceed the rate needed to replace water lost across the integument. For example, drinking rates in *A. campestris* inhabiting hypersaline media exceed water loss across the integument by 12-fold (Fig. 2b). This leads to the surprising result that larvae in saline water produce a greater volume of urine than those in dilute waters (compare Figs. 2a and 2b). The high rate of drinking in saline-water larvae is thought to serve two purposes, 1) the replacement of water to regulate body volume and 2) the ingestion of nutrients dissolved in the external medium (Bradley & Phillips, 1977b).

As a result of a rapid rate of drinking, larvae in hypersaline media are faced with a tremendous salt load. The remainder of this review deals principally with the osmoregulatory organs producing the excreta and the physiological mechanisms contributing to ionic regulation in the hemolymph.

II. The Site of Hyperosmotic Urine Formation

Beadle (1939) was the first to examine osmoregulation in saline-water larvae. By placing ligatures at various locations on larvae of *A. detritus*, he demonstrated that regulation was dependent upon organs in the posterior portion of the animal. Ramsay (1950) examined the osmotic concentration of the hemolymph and fluid in the midgut,

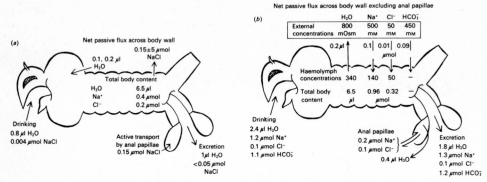

Fig. 2. Net exchanges per day via different body sites for 8 mg larvae of *A. campestris* in hyposmotic (a) or hyperosmotic (b) waters. Details of experimental procedures are given in Bradley and Phillips (1977b).

Malpighian tubules and rectum of *A. detritus*. He showed that the rectum was the site of hyperosmotic urine formation. Ramsay pointed out that the rectum of an obligatorily freshwater larva, *A. aegypti*, which produces a hyposmotic urine, is composed of a single segment, while the rectum of saline-water larvae is divided into histologically distinct segments (Fig. 3). Ultrastructural studies conducted on the saline-water larvae of *A. campestris* demonstrated that each segment of the rectum is composed of a single, morphologically distinct cell type (Meredith & Phillips, 1973). The cells of the anterior rectal segment were found to resemble the single cell type comprising the rectal epithelium of *A. aegypti* (Fig. 3). These cells have well-developed apical and basal membrane infoldings. Mitochondria are distributed evenly throughout the cytoplasm. The cells of the posterior rectal segment are approximately twice as thick as those of the anterior segment. The apical folds extend across ca. 60% of the cell cytoplasm. The majority of the mitochondria in the cells of the posterior rectum are associated with the apical membrane infoldings. Both anterior and posterior rectal cells have relatively narrow, straight intercellular channels in contrast to the highly folded and intermittently widened intercellular spaces which are associated with fluid reabsorption in terrestrial insect recta (reviewed by Phillips, 1980). On the basis of Ramsay's measurements of osmotic concentration and the ultrastructural characteristics of the larval recta, Meredith & Phillips (1973) proposed that the production of a hyperosmotic urine in saline-water mosquito larvae was by salt secretion rather than by water resorption. They proposed that hyperosmotic secretion occurred in the posterior rectal segment since the cells in this region are unique to saline-water species and their ultrastructure is indicative of a very active role in ion transport.

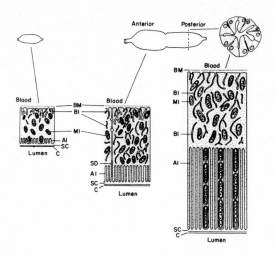

Fig. 3. Morphology of rectum of freshwater mosquito (Left) (*A. aegypti*) and saline-water mosquito (Right) (*A. campestris*) larvae. Based on electron micrographs by Meredith and Phillips (1973) and drawn by Wall and Oschman (1975). Drawn approximately to scale. BM: basement membrane, BI: basal infold, MI: mitochondrion, AI: apical infold, SD: septate junction, SC: subcuticular space, C: cuticle.

Bradley & Phillips (1975) were able to demonstrate that hyperosmotic fluid secretion is indeed the mechanism of formation of a concentrated excreta in saline-water mosquitoes. Larvae of *A. taeniorhynchus* were ligated anterior to the rectum and at the anal segment. These ligations prevented fluid movement into or out of the rectum except by net fluid movement across the rectal wall. The terminal segments containing the ligated rectum were suspended from the surface of saline solutions by means of the respiratory siphon, thereby assuring a supply of oxygen via the normal tracheal connections. The integument was torn and the preparation was placed in saline of known composition. Under such conditions Bradley & Phillips (1975) observed that the rectum fills with a hyperosmotic secretion. This phenomenon of hyperosmotic fluid secretion is clearly distinct from the mechanism of concentrated excreta formation in terrestrial insects, which involves resorption of a hyposmotic fluid (Phillips, 1980).

III. Mechanisms of Ion Transport

A. The Rectum

The role played by the rectum in the ionic regulation of the hemolymph was examined by Bradley & Phillips (1975, 1977a,b,c). They found that the concentrations of Na^+, Mg^{++} and Cl^- secreted into the rectum were higher than those found in the hemolymph and tended to resemble the concentrations of these ions in the external medium to which the animal had been acclimated (Fig. 4). The concentrations of Na^+ and Cl^- in the secretion were sufficient to account fully for the elimination of these ions in hypersaline media. The concentraton of potassium in the rectal secretion was always greater than in either the hemolymph or the external medium (Fig. 4). Excretion of rectal fluid therefore leads to a net potassium loss which presumably is balanced by potassium intake, perhaps in the food. Sulphate ions occur in high concentrations in waters inhabited by saline-water larvae. In seawater the sulphate concentration is 28mM while in inland waters larvae have been found in waters containing 300mM $SO_4^=$ (Kiceniuk & Phillips, 1974). Surprisingly, sulphate is not actively transported across the rectal wall in such larvae (Bradley & Phillips,

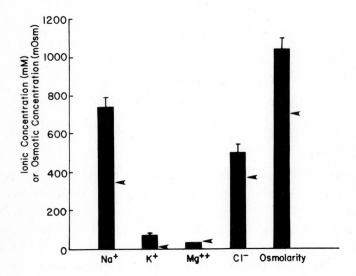

Fig. 4. Concentrations of ions in the rectal secretion (bars) compared to concentrations of the same ions in the external medium (arrowheads). Based on data from Bradley and Phillips (1977a).

1977a).   Instead,   the Malpighian tubules   are the major site of
excretion for this ion (see section B).

Bradley & Phillips (1977c) measured transepithelial electrical
potentials in the recta of *A. taeniorhynchus* during periods of fluid
secretion.   The anterior (-13 mV lumen relative to hemolymph) and
posterior (+11 mV) rectal segments have potentials of opposite
polarity, in keeping with their presumed differences in physiological
function.   These measurements permitted estimates to be made of the
electrochemical gradients for each ion.   The authors found that Na^+,
K^+, Cl^- and Mg^{++} were all actively transported across the rectal
wall.   On the basis of transmembrane electrical potentials and
presumed intracellular ion concentrations, a model of the locations of
active and passive ion movements across membranes of the posterior
rectum was proposed (Fig. 5).

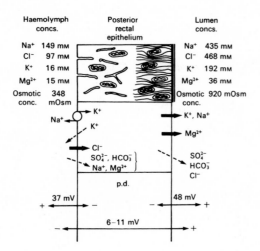

Fig. 5.   A model by Bradley and Phillips (1975) for the cellular
location of ion transport processes in the posterior rectum of
A. taeniorhynchus reared in seawater.   Solid arrows indicate
active transport and broken arrows depict passive movements.
The arrows indicate expected transport sites if the ionic
gradients described originate solely by hyperosmotic secretion
in the posterior rectal segment.

Strange (1983) has conducted detailed studies of the mechanism of
HCO_3^- regulation in saline-water mosquito larvae.   For this purpose he
used larvae of *Aedes dorsalis*, a species capable of surviving in

hypersaline alkaline waters. Strange et al (1982) demonstrated that HCO_3^- secretion in the rectum of *A. dorsalis* larvae acclimated to hypersaline waters was an active transport process. Lumen to hemolymph HCO_3^- and $CO_3^=$ gradients of 21:1 and 241:1, respectively, are generated by the rectal epithelium against a transepithelial potential of -25 mV (lumen negative).

To examine HCO_3^- transport in further detail, an *in vitro* microperfused rectal preparation was developed (Fig. 6)(Strange and Phillips, 1984). Net transport of HCO_3^-, $CO_3^=$ and CO_2 (termed net CO_2 transport) $J_{net}^{CO_2}$ was measured by microcalorimetry. $J_{net}^{CO_2}$ was unaffected in perfused salt glands by bilateral Na^+ or K^+ and serosal Cl^- substitutions, or by serosal addition of 1.0 mM ouabain, 2.0 mM amiloride or 0.5 mM SITS. Removal of luminal Cl^- inhibited $J_{net}^{CO_2}$ by 80%, while serosal addition of 1.0 mM acetazolamide or 0.5 mM DIDS inhibited $J_{net}^{CO_2}$ by 80% and 40%, respectively.

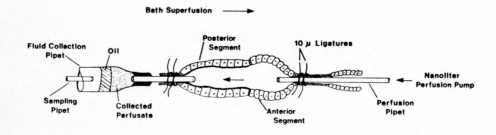

Fig. 6. A diagram illustrating the perfused rectal preparation (from Strange et al., 1983).

Separate perfusion of the anterior and posterior rectal segments demonstrated that the anterior rectum was the site of CO_2 secretion in the microperfused rectum (Strange et al, 1984). Net Cl^- reabsorption in the anterior segment was measured by electron microprobe analysis and was equivalent to the rate of CO_2 secretion. In addition, Cl^- reabsorption in the anterior segment was completely inhibited by bilaterally replacing CO_2 and HCO_3^- with a phosphate or HEPES buffered saline. These studies provide strong quantitative evidence for the presence of a 1:1 Cl^-/HCO_3^- exchange mechanism located in the anterior rectal segment.

The cellular mechanisms of anterior salt gland HCO_3^- and Cl^- transport were examined using ion and voltage sensitive microelectrodes in conjunction with a microperfused preparation of the anterior rectal segment. The preparation allowed complete changes in serosal and mucosal bathing media to be made in less than 5-10 seconds (Strange & Phillips, in prep.). Addition of DIDS or acetazolamide to, or removal of CO_2 and HCO_3^- from the serosal bath caused large, 20-50 mV hyperpolarizations of the electrical potential across the apical membrane (V_a) and had little effect on the potential across the basolateral membranes (V_{bl}). Rapid changes in luminal Cl^- concentration altered V_a in rapid, step-wise manner. The slope of the relationship between V_a and luminal Cl^- activity was 42.2 mV per decalog change in the luminal activity of Cl^- (r=0.992). Intracellular Cl^- activity was 23.5 mM and was approximately 10mM lower than that predicted for a passive distribution at the apical membrane. Changes in serosal Cl^- concentration had no effect on V_{bl}, indicating an electrically silent basolateral Cl^- exit step. Intracellular pH in anterior rectal cells was 7.67 and the calculated intracellular activity of HCO_3^- was 14.4mM. These data show that under control conditions, HCO_3^- enters the anterior rectal cell by an active mechanism against an electrochemical gradient of 77.1 mV and exits the cell at the apical membrane down a favorable electrochemical gradient of 27.6 mV. Based on these results a tentative cellular model has been proposed in which Cl^- enters the apical membrane of the anterior rectal cells by a passive, electrodiffusive movement through a Cl^--selective channel. HCO_3^- exit from the cell occurs by an active or passive electrogenic transport mechanism. The electrically silent nature of basolateral Cl^- exit and HCO_3^- entry, and the effects of serosal addition of the Cl^-/HCO_3^- exchange inhibitor DIDS on $J_{net}^{CO_2}$ and transepithelial voltage strongly suggest that the basolateral membrane is the site of a direct coupling between Cl^- and HCO_3^- movements via a Cl^-/HCO_3^- exchange mechanism (Fig. 7).

At present, the precise functions of the individual rectal segments are unclear. Bradley & Phillips (1977c) examined segmental function in *A. taeniorhynchus* by tying ligatures around isolated recta to examine the anterior and posterior rectal segments in isolation. Each segment was incubated in artificial hemolymph for two hours and the recta were examined for swelling due to fluid secretion. The posterior segment swelled slightly with fluid which was hyperosmotic to the artificial hemolymph. No swelling and fluid secretion were apparent in anterior rectal segments. The very small samples of fluid which could be removed from the lumen of the anterior segment were hyposmotic to the bathing medium. These results in conjunction with

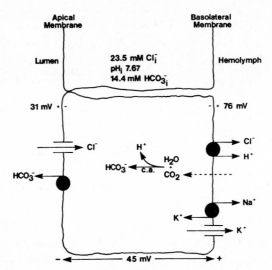

Fig. 7. A model by Strange (1983) illustrating proposed HCO_3^- and Cl^- entry and exit steps in anterior rectal cells of *A. dorsalis*.

the ultrastructural studies of Meredith and Phillips (1973) (see above) led Bradley & Phillips (1977c) to hypothesize that the posterior rectum was the site of hyperosmotic fluid secretion while the anterior rectum was involved in selective reabsorption of inorganic and organic solutes.

Strange et al (1984) examined this hypothesis using microcannulated rectal segments from *A. dorsalis*. They found that both rectal segments are capable of hyperosmotic fluid secretion. The posterior and anterior rectal segments in *A. dorsalis* account for 75% and 25%, respectively, of the fluid secretion observed in the whole rectum. Analysis of the composition of the secretion demonstrated that both segments are capable of excreting Na^+, Cl^- and HCO_3^-. Based on these results and the results of studies in which the effects of serosal ion substitutions on rectal secretion were examined, Strange et al. (1984) have suggested that both segments normally secrete a hyperosmotic, NaCl-rich fluid, a secretion driven by coupled NaCl transport. Approximately 75% of this secretion occurs in the posterior rectum. Once this fluid enters the rectal lumen, its composition is modified by ion exchange and reabsorptive processes which are dependent upon the ionic regulation needs of the animal. The only such process characterized to date is the Cl^-/HCO_3^- exchange found in the anterior rectal segment of larvae from bicarbonate-rich lakes.

B. The Malpighian Tubules

Ramsay (1951) observed that the osmotic concentration of fluid in the Malpighian tubules of mosquito larvae was always essentially isosmotic to the hemolymph. Analysis of the principal monovalent ions in the fluid indicated that mosquito larvae, like most insects, produce a primary urine rich in K^+ and Cl^- and low in Na^+. Larvae reared in seawater produce a similar fluid despite the heavy influx of Na^+ which they experience. As a result, the secretion of primary urine by the Malpighian tubules in saline-water mosquito larvae contributes little to the regulation of osmotic and Na^+ concentrations in the hemolymph.

When secreting slowly, the Malpighian tubules of female, adult *A. taeniorhynchus* also produce a fluid with a high K^+ to Na^+ ratio (Maddrell & Phillips, 1978). Following the blood meal, however, the Malpighian tubules exhibit diuresis and secrete a large volume of Na^+-rich fluid. This capacity to produce a more Na^+-rich urine during postprandial diuresis is also observed in the hemipteran bloodsucking bug *Rhodnius prolixus*. The physiological mechanisms determining the K^+ to Na^+ ratios in the primary urine have not been examined in mosquitoes, but Maddrell (1977) has proposed that increased Na^+ secretion during diuresis is the result of increased Na^+ permeability in the basal plasma membrane of the Malpighian tubule. It seems therefore that the Malpighian tubules of the adults do contribute to Na^+ excretion as an adaptation to bloodfeeding. Those of the larvae do not, however, even in saline-water species living in Na^+-rich waters.

In addition to the above systems for the transport of monovalent ions, the Malpighian tubules of saline-water mosquito larvae show interesting and unusual properties with regard to the transport of divalent cations. Maddrell and Phillips (1975) found that the Malpighian tubules of *A. campestris* and *A. taeniorhynchus* were able to transport $SO_4^=$ against the electrochemical gradient for this ion. Sulphate tranport displayed Michaelis-Menton kinetics with a Km of 10mM and a V_{max} of 50pmol min-1 tubule^{-1} (Fig. 8a). In a later study, these authors found that sulphate transport in the tubules was inducible (Maddrell & Phillips, 1978). Tubules from animals reared in sulphate-free artificial seawater showed very low levels of sulphate transport *in vitro*. Animals reared in sulphate-enriched seawater showed increased rates of sulphate tranport in proportion to the level of this ion in the rearing medium (Fig. 8b). If larvae reared in sulphate-free seawater were placed in seawater augmented with 89mM

Na_2SO_4, increased transport of sulphate by the Malpighian tubules was demonstrable within 8 h of transfer. Augmentation of transport occurred through an increase in V_{max} with no change in K_t. These results suggest that the presence of sulphate in the external medium results in the synthesis and insertion of additional tranport "pumps" into the Malpighian tubule membranes. A precise knowledge of 1) the form of the stimulus within the larva 2) the mechanism of induction, and 3) whether the observed transport is primary or is coupled to the movement of other ions, remains unknown.

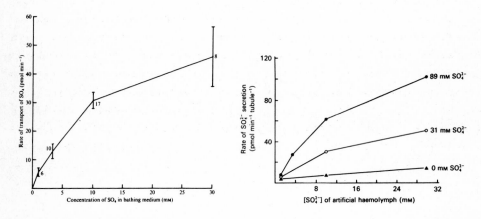

Fig. 8. (a) The dependence of the rate of secretion of sulphate by single isolaed Malpighian tubules of A. *taeniorhynchus* on the concentration of these ions in the bathing medium. (b) The influence of external $SO_4^=$ levels during early development of A. *taeniorhynchus* larvae on the capacity of Malpighian tubules to secrete this anion *in vitro*.

The Malpighian tubules of A. *campestris* possess a mechanism for the active transport of magnesium ions (Phillips & Maddrell, 1974). Magnesium transport exhibits saturation kinetics with a Kt of 2.5mM and V_{max} of 15pmol min^{-1} $tubule^{-1}$. Recently, Ng and Phillips (unpublished observations) have found an inducible mechanism for Mg^{++} transport in the Malpighian tubules of A. *dorsalis*, using the approach outlined above for sulphate.

Bradley et al. (1982) conducted an ultrastructural study of the Malpighian tubules from larvae of A. *taeniorhynchus* reared in seawater. They found that the tubules were composed of two cell types: the large primary cells thought to be the major site of fluid transport, and the smaller stellate cells whose physiological function is unknown. These authors found that the cell types in the tubules of the saline-water mosquito larvae *A. taeniorhynchus* are identical to those found in (*A. aegypti*), a freshwater species. This indicates that, unlike the larval recta of saline-water and freshwater mosquitoes which differ in cell types, the unique capacity of the tubules of saline-water mosquito larvae to transport Mg^{++} and $SO_4^=$ is not associated with the presence of an additional cell type.

SUMMARY

By drinking the external medium, saline-water mosquito larvae replace water lost by osmosis across the cuticle and provide excess water for urine production. Ions accompanying the ingested water are excreted in the form of a urine strongly hyperosmotic to the hemolymph and slightly hyperosmotic to the external medium. Figure 9 illustrates demonstrated pathways of ion transport and their probable location. The Malpighian tubules produce an isosmotic KCl-rich secretion. These organs also have inducible active transport mechanism for $SO_4^=$ and Mg^{++}. The Malpighian tubules are the only site

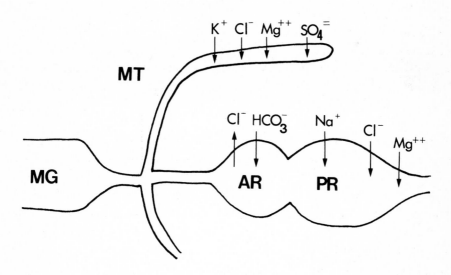

Fig. 9. Demonstrated pathways of ion transport and their probable locations. See text for further explanation.

of active sulphate extrusion since this anion is not actively transported in the rectum.

Formation of a concentrated urine occurs through secretion of a hyperosmotic fluid in the rectum. Na^+, Mg^{++}, K^-, and Cl^- are all actively transported from the hemolymph into the rectal lumen. As described above, secretion of a hyperosmotic fluid may occur in both the anterior and posterior segmens, but the majority of such transport occurs in the posterior segment. In larvae reared in bicarbonate-rich media, modification of the fluid in the rectal lumen has been shown to occur by means of Cl^-/HCO_3^- exchange. This exchange, which serves in pH regulation and the retention of Cl^- which would otherwise be lost in the urine, is found in the anterior rectal segment.

REFERENCES

Asakura, K (1980) The anal portion as a salt secreting organ in a seawater mosquito larva, *Aedes togoi* Theobald. J. Comp. Physiol. 138 : 346-362.

Beadle LC (1939) Regulation of the hemolymph in saline-water mosquito larva *Aedes detritus* Edw. J. exp. Biol. 16 : 346-362.

Bradley TJ and Phillips JE (1975) The secretion of hyperosmotic fluid by the rectum of a saline-water mosquito larva, *Aedes taeniorhynchus*. J. exp. Biol. 63 : 331-342.

Bradley TJ and Phillips JE (1977a) Regulation of rectal secretion in saline-water mosquito larvae living in waters of diverse ioinic composition. J. exp. Biol. 66 : 83-96.

Bradley TJ and Phillips JE (1977b) The effect of external salinity on drinking rate and rectal secretion in the larvae of the saline-water mosquito, *Aedes taeniorhynchus*. J. exp. Biol. 66 : 97-110.

Bradley TJ and Phillips JE (1977c) The location and mechanism of hyperosmotic fluid secretion in the rectum of the saline-water mosquito larvae, *Aedes taeniorhynchus*. J. exp. Biol. 66 : 111-126.

Bradley TJ, Stuart AM and Satir P (1982) The ultrastructure of the larval Malpighian tubules of the saline-water mosquito, *Aedes taeniorhynchus*. Tissue & Cell 14(4) : 759-773.

Kiceniuk JW and Phillips JE (1974) Magnesium regulation in mosquito larvae, *Aedes campestris*, living in waters of high $MgSO_4$ content. J. exp. Biol. 61 : 749-760.

Maddrell SHP and Phillips JE (1975) Active transport of sulphate ions by the Malpighian tubules of the larvae of the mosquito *Aedes campestris*. J. exp. Biol. 62 : 367-378.

Maddrell SHP and Phillips JE (1978) Induction of sulphate transport and hormonal control of fluid secretion by Malpighian tubules of larvae of the mosquito *Aedes taeniorhynchus*. J. exp. Biol. 72 : 181-202.

Maddrell SHP (1977) Insect Malpighian tubules. In : Gupta BL, Moreton RB, Oschman JL and Wall BJ (eds) Transport of ions and water in animals. Academic Press, London p 541.

Meredith J and Phillips JE (1973) Rectal ultrastructure in salt and freshwater mosquito larvae in relation to physiological state. Z. Zellforsch. Mikr. Anat. 138 : 1-22.

Nayar JK and Sauerman DMJr (1974) Osmoregulation in larvae of the salt-marsh mosquito, *Aedes taeniorhynchus*. Ent. exp. appl. 17 : 367-380.

Nicholson SW (1972) Osmoregulation in larvae of the New Zealand salt-water mosquito *Opifex fuscus* Hutton. J. of Entomol. 47 : 101-108.

Nicholson SW and Leader JP (1974) The permeability to water of the cuticle of the larva of *Opifex fuscus* (Hutton) (Diptera, culicidae). J. exp. Biol. 60 : 593-604.

Phillips JE (1980) Epithelial transport and control in recta of terrestrial insects. In : Locke ML and Smith DS (eds.) Insect biology in the future. Academic Press, NY. p 145.

Phillips JE, Bradley TJ and Maddrell SHP (1978) Mechanisms of osmotic and ionic regulation in saline-water mosquito larvae. In : Schmidt-Nielsen K, Bolis L and Maddrell SHP (eds.) Comparative physiology : water, ions and fluid mechanics. Cambridge University Press, Cambridge. p 151.

Phillips JE and Maddrell SHP (1974) Active transport of magnesium by the Malpighian tubules of the larvae of the mosquito, *Aedes campestris*. J. exp. Biol. 61 : 761-771.

Ramsay JA (1950) Osmotic regulation in mosquito larvae. J. exp. Biol. 27 : 145-157.

Ramsay JA (1951) Osmoregulation in mosquito larvae: the role of the Malpighian tubules. J. exp. Biol. 28 : 62-73.

Sheplay AW and Bradley TJ (1982) A comparative study of magnesium sulphate tolerance in saline-water mosquito larvae. J. Insect Physiol. 28(7) : 641-646.

Strange K (1983) Cellular mechanism of bicarbonate regulation and excretion in an insect inhabiting extremes of alkalinity. PhD Thesis, University of British Columbia, Vancouver. 218 pp.

Strange K and Phillips JE (1984) Mechanisms of CO2 transport in the microperfused rectal salt gland of *Aedes dorsalis*. I. Ionic requirements of CO_2 secretion. Am J. Physiol. (in press).

Strange K, Phillips JE and Quamme GA (1982) Active HCO_3^- secretion in the rectal salt gland of a mosquito larva inhabiting $NaHCO_3-CO_3$ lakes. J. exp. Biol. 101 : 171-186.

Strange K, Phillips JE and Quamme GA (1984) Mechanisms of CO_2 transport in the microperfused rectal salt gland of *Aedes dorsalis*. II. Site of Cl^-/HCO_3^- exchange and function of anterior and posterior salt gland segments. Am. J. Physiol. (in press)

Wall BJ and Oschman JL (1975) Structure and function of the rectum in insects. In : Wessing A (ed) Fortschritte der Zoologie 23 : 192-222.

ACKNOWLEDGEMENTS

Original research by the authors was supported by NIH AI 17736 and NSF PCM 8215420 to T.J.B. and by a grant from NSERC of Canada to J.E.P.

CHLORIDE SECRETION BY THE CHLORIDE CELLS
OF THE ISOLATED OPERCULAR EPITHELIUM OF MARINE FISH

J.A. ZADUNAISKY

I. INTRODUCTION AND BRIEF HISTORY OF THE CHLORIDE CELL

Sea water teleosts drink a highly concentrated NaCl solution and utilize the
secretory epithelium of their gills to keep a homeostatic concentration of these
electrolytes in their blood, the kidney having a secondary role (Homer Smith, 1930).
The secretory epithelium of the gills consists mainly of the chloride cells, which
were predicted and discovered by A. Keys in the 1930's (Keys, 1931; Keys and Wilmer,
1932). By means of vascular perfusion in an isolated gill preparation Keys demonstrated a reduction in salts in the perfusate consistent with an increase in the medium
bathing the outside of the gills. The anatomical work with Wilmer demonstrated the
presence of large, complex "secretory like" cells in the gills of eels that degener-
ated or were reduced in number and size during adaptation to fresh water. In spite
of opinion to the contrary (Bevelander, 1935; 1936) modern physiological and anatomical methods have confirmed Ancel Keys findings. However, the actual driving forces
responsible for the movements of salts could not be determined conclusively until
very recently. The reason for the delay in a good biophysical demonstration of the
chloride secretion of these cells was the need for an in vitro preparation containing
chloride cells that permitted the application of the Ussing methodology (Ussing and
Zehran, 1951) to a chloride cell rich flat epithelial membrane. This type of membrane rich in chloride cells are found in the epithelium lining the opercular flap of
teleosts (Burns and Copeland, 1950). The opercular epithelium can be dissected,
mounted as a membrane and actual ionic fluxes, electrical potentials and short circuit current determined to explain the function of these cells (Degnan, Karnaky, and
Zadunaisky, 1977). In this presentation the information obtained up to now is presented focusing on the opercular epithelium of Fundulus heteroclitus. Reference to
other more recently described opercular epithelia are presented also in the text.

II. THE CHLORIDE CELL STUDIED IN THE OPERCULAR EPITHELIUM

The epithelium contains numerous chloride cells, about 40,000 per cm^2; these are found together with pavement cells, mucous cells and non-differentiated cells. The secretory function has been demonstrated to reside in the chloride cells for the following reasons: (a) there is high statistical correlation between the number of cells in the epithelium and the short circuit current due to chloride transport in F. heteroclitus(Karnaky et al.,1979); in F. grandis (Krasney and Zadunaisky,1978) and in the skin of Gallichtys mirabilis (Marshall and Nishioka, 1980);   (b) the number and size of chloride cells increases with adaptation to sea water in the opercular epithelium of Telapia (Foskett et. al., 1981) and the chloride current increases proportionally to number and size; (c) studies with the vibrating probe have demonstrated the presence of low resistance pathways that carry chloride current outward in the pyths of opercular epithelia of telapia(Scheffey et al., 1983). The chloride cells of the opercular epithelium are identical to the ones of the gill epithelium in anatomy, binding of labeled ouabain and response to sea water adaptation (See Zadunaisky, 1984).

A. Electrical potentials and ionic fluxes

The potential difference across the isolated opercular epithelia is oriented outside negative, as in the intact gills. The potentials range between 10 and 35 mV. The average short circuit current is 130 $\mu A/cm^2$ and the electrical d.c. resistance 173 ohms·cm^2 for Fundulus heteroclitus. Similar values have been found for F. grandis and Telapia, and somewhat lower values for Gallichtys mirabilis (see Zadunaisky, 1984). In all cases the short circuit current was identical to the net flux of chloride from blood side to sea water side. The values obtained ranged from 0.6 to 2.5 nA per chloride cell present in the opercular membranes. Flux values in F. heteroclitus operculii averaged 6 to 7 $\mu Eq/hr/cm^2$ for the net chloride flux towards the sea water electronegative side.

B. Influence of other ions and drugs on chloride secretion

Anoxia and poisoning with metabolic inhibitors produced a remarkable reduction in chloride current (Degnan et al., 1977). Chloride free solutions in both sides produced reduction to extremely low levels with recovery on readmission of chloride ions to the preparation. Sodium was required for the transport to occur, though it is not the driving force for the secretion. There was no net sodium flux across these preparations (Degnan, et al., 1977; see also review by Zadunaisky, 1984). Bicarbonate had great influence on the chloride secretory process and it was required in the blood side of the preparation. Optimal maximum values were observed with 16 mM bicarbonate in the medium, however higher concentrations produced even higher values for the chloride net flux. The change of pH produced by bicarbonate did not account for the effects of the ion itself. It is possible that either bicarbonate in the medium or free CO_2 in the cell is responsible for this effect.

Open circuit measurements of ion fluxes in F. heteroclitus opercular epithelia demonstrated that the ratio of chloride fluxes was approximately 9-10 times different than the one predicted by Ussing's flux equation (Ussing, 1949). Sodium passive fluxes on the other hand conformed to the predicted passive value confirming the absence of net sodium flux across the preparation (Degnan and Zadunaisky, 1979).

K ions are required in the inside bathing solution to maintain the current and its effects are extremely rapid (Degnan and Zadunaisky, 1980b). Ouabain, the inhib-itor of NaK ATPase has a drastic inhibitory effect on the chloride secretion; furo-semide, the specific blocker of chloride transport produced also inhibitory effects on the chloride current of the opercular membranes. Thiocyanate at high concentra-tion competed for the chloride site of transport. Diamox did not inhibit the chlor-ide current, it had a small but statistically significant stimulation of the short circuit current. Amiloride had no effects at low concentrations and a small (10%) reduction in current at high concentrations. All these effects were observed in F. heteroclitus membranes (Degnan et al., 1977). Most of these effects of other ions and drugs are directly related to the chloride transport functions of the chloride cell. Ouabain implies the presence of the sodium pump, that keeps the sodium level in the cell low by pumping sodium in the inward membrane towards the blood, in ex-change for K. Furosemide acts on the Na K Cl coupler most probably located at the entry step in the basolateral side. Bicarbonate is probably related to a Cl^-/CO_2H^- exchange, and sodium as well as potassium are required for the functions of the furosemide sensitive coupler at the chloride entry step as well as for the function of the sodium pump.

C. Opercular membranes other than F. heteroclitus

Following the pattern of the chloride secretion in F. heteroclitus opercular epithelium, developed in our laboratory, other preparations have been used with sim-ilar success. The most active is the opercular membrane of Telapia (Sarotherodon mossambicus) in which the net current is also equal to the chloride secretion, and in general behaves as the chloride cell rich operculum of Fundulus (Foskett et al., 1981). In Telapia, the chloride cells are reduced in number in fresh water and the adaptation to sea water coincides with increases in size and number of the chloride cells. The effects of prolactin (Foskett et al., 1981) and of cortisone have been studied in these membranes and are discussed further on. The skin of Gallichthys mirabilis also contain chloride cells and has been studied as a membrane (Marshall 1977). It produces potentials of the same order of magnitude of Fundulus and Telapia, however, shows smaller net chloride transport. The opercular epithelium of Fundulus grandis (Krasny and Zadunaisky, 1978) also has similar properties, with somewhat smaller numbers of chloride cells per cm^2 and in general with similar electrical and transport properties as the other species.

III. THE PARACELLULAR SHUNT PATHWAYS AND PASSIVE ION CONDUCTANCES

A detailed study of the paracellular pathway was undertaken by Degnan and
Zadunaisky (1980a). The reason for the study is that it can provide information as
to the ions that move through the cell and ions that move through the paracellular
shunt. It can also provide information on whether the passive ions move through the
same pathway or not.

It was found that Na conductance explained 50 percent of the total electrical
conductance and that when either influxes or effluxes of sodium were plotted against
total electrical conductance linear correlation was obtained. This eliminates sodium again as a source of current in the system and explains its passage through the
paracellular shunt pathway. Similar comparisons, of total electrical conductance
versus partial electrical conductances calculated from urea fluxes indicated no correlation at all.

Another manner in which the paracellular pathway was studied was to increase or
decrease the driving force on the ions by voltage clamping the potential 25 mV above
or below the zero value provided by the short circuiting technique. Under these conditions the values of the unidirectional fluxes for passive ions moving through the
paracellular pathway could be predicted on the basis of the flux equations (see
Degnan and Zadunaisky, 1980a). It was found that at the two levels of voltages utilized, the predicted versus the observed movements of sodium were identical, indicating their movement through the paracellular pathway. Under these conditions two
agents that modify sodium cellular movements, amiloride and amphotericin b were
tested and neither had effects on the sodium fluxes, indicating again that the paracellular nature of the sodium pathway. Furthermore, total electrical conductance
was not affected by agents that are known to affect cellular conductance.

Still another test for sodium conductance was performed utilizing TAP (Moreno,
1975). This agent has the property of blocking sodium paracellular shunts. In fact,
it had a pronounced effect on the electrical conducance of the opercular membranes.
However, the ratio sodium conductance/total electrical conductance remained around
the value of 0.50 indicating that in spite of a reduction of both tissue and sodium
conductance of 70%, the portion of ionic conductance for sodium was unchanged. No
effects of TAP on urea conductance were found.

Still another test for the paracellular pathways consisted in eliminating Na
from the bathing solutions and determining partial conductances. It was found that
sodium affects the total electrical conductance. However, the short circuit current
was reduced dramatically indicating the great sodium dependence of the chloride
transport in this system.

Similar types of experiments as referred above for the sodium pathways, were
performed for the passive chloride flux, that is in the direction sea water side to

blood side. It was observed that the total electrical conductance is
not statistically correlated with passive chloride conductance, and in
this sense the passive chloride flux behaves like the urea flux. Second, the effect of clamping to the high and low voltages indicated that
the inward component was a passive chloride flux moving through a paracellular pathway. However, TAP had no effect on the passive chloride
flux, and this observation could indicate two different pathways in the
paracellular shunt: one for sodium and a different one for chloride and
urea.

The results of these studies of Degnan and Zadunaisky (1980a) permit then the conclusion that sodium moves passibly across the epithelium
and not through the cells, and that predicted Na/Na exchanges can be
eliminated in studies of a more rigorous nature in isolated preparations.
The general conclusion of the study of the paracellular pathways then
are (1) the passive chloride flux and both sodium unidirectional fluxes
occur through the paracellular pathways and not through the chloride
cells. (2) There are different paracellular pathways for sodium and for
chloride and urea. (3) The results do not agree with predictions of a
Na/Na and Cl/Cl exchange as proposed by Motet et al. (1966) and the
Na/K exchanges proposed by Maetz (1969). Neither agree with the results on active sodium efflux proposed by Potts and Eddy (1973) across
the sea water adapted flounder gill.

A. Probable nature of the signal for secretion in changing from sea
    water to fresh water

The rapid adaptation to low sodium chloride environments of the intact fish, could be explained if the total electrical conductance or
in simpler words the permeability of the paracellular pathway was sensitive to sodium concentration in the sea water side. When the sodium
is rapidly reduced on that side, the total electrical conductance of
the chloride cell rich preparations is drastically reduced at the ex-
pense of a reduction of sodium unidirectional movement through the paracellular pathway. It is possible that there is a special region in the
gill that detects sodium activity in the outside medium and reduces or
increases sodium conductance when sodium is reduced or increased. After
this rapid shut off of the paracellular pathway, then the epithelium or
extrapolating, the secretory gill epithelium will become more tight and
will not permit the movement down the gradient of salts when exposed
to fresh water. In time, this rapid mechanism would be superseded by
the action of prolactin that has a slower time of action and has been
shown to produce similar effects on total gill permeability or on the

permeability of isolated opercular membranes (Mayer Gostan and Zadunaisky, 1978; Foskett et al., 1982).

IV. WATER MOVEMENTS ACROSS THE OPERCULAR EPITHELIUM

The net movement of chloride must be accompanied by water across the epithelium. In isolated opercular membranes of _Fundulus heteroclitus_ experiments were performed by Brown and Zadunaisky (1982) that indicate that there is a movement of 7.2 $\mu l/hr/cm^2$ of water from the blood to the sea water side when the preparation was bathed in Ringer's solution in both sides. Tissues with higher electrical resistances, above 170 ohms·cm^2 produced this flow, while more permeable membranes below the indicated resistance showed little or no flow. The addition of isoproterenol that increases the chloride secretion in _Fundulus operculi_ (see action on catecholamines) produced a remarkable increase in water flow.

V. HORMONAL EFFECTS AND SPECIFIC RECEPTORS IN THE CHLORIDE CELL

A. Catecholamines

The effect of catecholamines on the intact gills and in perfused gill preparations is well known. However, it is difficult to separate the circulatory or vascular effects from the action at the level of secretion in the gill. In isolated operculii of F. heteroclitus catecholamines have a pronounced effect. The addition of beta agonists such as isoproterenol produces a rapid and sustained increase in chloride current at concentrations below 10^{-7} or with maximal effects at $10^{-5}M$ (Degnan et al., 1977; Degnan and Zadunaisky, 1979). The stimulatory action of isoproterenol is inhibited by propranolol.

The addition of alpha agonists, such as noradrenaline produces on the other hand a rapid inhibition of the chloride secretion. The effects occurs also at very low concentrations and in this case inhibition by phentolamine can be easily detected. The increases in current associated with addition of beta agonists is accompanied by an increase in cyclic AMP in the tissues. Mendelsohn et al. (1981) showed a significant increase in cyclic AMP of isolated incubated operculii of F. heteroclitus during the action of isoproterenol.

Adrenaline has a biphasic effect, first inhibition and then stimulation as could be expected from the agonistic beta and alpha stimulation of this catecholamine.

For both beta and alpha agonists the chloride net flux increases
or decreases with the current, maintaining the relationship of short
circuit current and chloride net flux (Degnan and Zadunaisky, 1979).

In the case of F. grandis the beta agonist effect was not that remarkable. The findings in the isolated opercular membranes permit the
conclusion that catecholamines have a direct regulatory effects on the
chloride secretory process independently of their effects on the vascular system. Some of the effects in perfused gills found by Shuttleworth (1978) can be explained by these observations in opercular membranes.

Similar actions of catecholamines are also found in Gallichytis
mirabilis skin (Marshall and Bern, 1979) and in the operculum of Telapia
(Foskett et al., 1982) where actions of adrenaline has mostly an inhibitory effect on the chloride current.

B. Acetylcholine

The addition of acetylcholine to the bathing media produces inhibition of the chloride current and chloride net flux in F. heteroclitus
opercular epithelia (Rowing and Zadunaisky, 1978). Acetylcholine is
most effective when applied on the serosal or blood side, and its effects
are not affected by nicotin. Muscarine instead mimics the action of
acetylcholine and the muscarinic antagonist homatropine produced a
shift in the acetylcholine dose response curve, indicative of competitive inhibition. Nicotine did not speed the action of acetylcholine
and after nicotine the preparations were still responsive to carbachol,
eliminating a nicotine receptor. Escerine produced no changes in the
responses to acetylcholine. The reduced short circuit current under
acetylcholine was concurrent with a reduction in chloride net flux
accounting for the reduced secretion.

C. Prolactin

Prolactin is well known to control osmoregulation in the gills
(Maetz and Burnicin review of 1975) in intact fish. In the isolated
opercular preparations prolactin produces inhibition of chloride secretion (Mayer-Gostan and Zadunaisky, 1978) and its site of action must
be on or around the chloride cell. The effects were obtained after
injecting with prolactin sea water adapted F. heteroclitus. No immediate effects were observed when prolactin was added to the solutions
bathing operculii of non-injected specimens. Therefore this important
hormone most probably accounts for the slow adaptation to fresh water
but not to the rapid capacity of euryhiline fish to change from high

to low salinities. As indicated above the rapid adaptive mechanism
most probably resides directly in the secretory epithelium. In a detailed study of the effects of prolactin on Telapia (Foskett et al.,
1982) it was found that injections of prolactin decreased both the
chloride current and the conductance in a dose dependent manner in isolated epithelia dissected from such specimens. Prolactin reduced the
number of chloride cells as well as inhibited reversibly the active
pathways or ionic conductances of the remaining cells.

D. Cortisol

     Cortisol is the most important corticoid in fish (Henderson et al.,
1970) and has been implicated for a long time in sea water adaptation.
The mechanisms of action of cortisol consists in a direct increase on
the number of chloride cells. In Telapia (Foskett et al., 1981) the
cell density in the opercular epithelium increases 2.5 times when compared to specimens not injected with cortisol. This increase in population of cells is similar to the increase produced by sea water adaptation of fresh water adapted specimens. However, the electrophysiological and transport properties do not go hand in hand. Cortisol treatment in fresh water specimens did not produce increases in the short-
circuit current or Cl secretion while sea water adaptation did. Most
probably, then, the exposure to sea water is needed to activate the
cell and make them transport chloride. Either the signal sits on the
secretory epithelium that is able to detect sodium concentrations, as
mentioned above, or the mechanism of sea water drinking and secretion
has to be started in the injected fresh water adapted fish before dissection and isolation of the operculum.

E. Glucagon and vasoactive intestinal polypeptide

     Agents that increase cyclic AMP levels in tissues were tested in
isolated opercular membranes of Telapia (Foskett et al., 1982). Glucagon at 10^{-9}M had stimulatory effects on the chloride current and was
potentiated by previous treatment with phosphodiaesterase inhibitors.
The potent agent VIP also produced increases in chloride current in
these epithelia rich in chloride cells from sea water adapted specimens
and was also potentiated by inhibitors of phosphodiesterase. Tissue
conductance was affected suggesting changes in membranes permeability
or a change in a paracellular pathway. These agents could have important effects in intact fish.

F. Urotensin I and II

The neurosecretory peptide urotensin I stimulates in vitro the
chloride current across the isolated skin of Gallichthys mirabilis
(Marshall and Bern, 1979). Urotensin I also reversed the previous inhibition produced by epinephrine. Urotensin II had inhibitory effects
and the authors concluded that there might be an antagonism between the
action of Urotensin I and Urotensin II and epinephrine.

VI. BASIS FOR A MODEL OF CHLORIDE SECRETION BY THE CHLORIDE CELLS

A comprehensive model for the chloride cell should include available evidence as well as predict some of its behavior. An attempt at
this stage lacks a fundamental piece of information, already available
in other chloride secretory cells and that is the intracellular activity of the main ions involved in the osmoregulatory process. However,
with the available evidence obtained in in vitro opercular preparations
and the understanding of how chloride secretion occurs in systems such
as the corneal epithelium (Zadunaisky, 1966, 1978, 1982) it is possible
to present a model for the chloride cell.

A. Experimental basis for the model of the chloride cell

The evidence obtained so far is the following: (a) there is a net
transport of chloride from basolateral to cript or apical side of the
epithelium, the chloride ions carrying all the current across the tissue in all the species examined, (b) there is no net movement of sodium across the epithelial cells, (c) nevertheless, the chloride active
secretion is sodium dependent, (d) the NaK pump inhibitor ouabain stops
the active chloride transport, (e) loop diuretics such as furosemide
inhibit profoundly the chloride active transport (f) activation of beta
receptors or addition of cyclic AMP stimulate Cl^- transport as in many
of the other secretory epithelia that show the same characteristics,
(g) NaK ATPase has been detected by means of histochemical techniques
or labeled ouabain binding to be located on the basolateral side of
the chloride cells of the gills or of the opercular epithelium (h)
amiloride has practically no effect on these preparations indicating
that sodium is not moving through the cellular pathway, (i) inhibitors
of cyclic AMP phosphodiesterase produce increases in chloride current
very similar to actions found in other chloride transporting epithelia
(j) thiocyanate a typical competitor of chloride transport has a very
clear inhibitory effect on the chloride cells (k) SITS has an inhibitory effect on chloride current of the opercular epithelium, which

points to the existence of a coupler such as the one found in other chloride transporting epithelial cells, (1) the basolateral side is extremely sensitive to reduction of K concentration indicating both the need for K for the operation of the NaK pump on that side and the probably requirement of K for Cl entry into the cell through a Cl/Na/K coupler.

All the above reasons permit the following interpretation. The NaK pump in the basolateral side of the chloride cells maintains the Na activity low inside the cytoplasm. The entry step of chloride from the blood or basolateral side is accomplished by a coupler that utilizes the sodium gradient and requires K. This side is sensitive to furosemide and SITS-like compounds. The apical side of the cell consists in a barrier were chloride channels would permit the passage by diffusion or facilitated diffusion of the chloride ions present in the cell at a higher activity than the one predicted by the electrochemical gradient. Sodium moves passibly through the paracellular pathway and chloride also utilizes a passive paracellular for the movement from sea water to blood, but this pathway as explained in the section on the paracellular pathways, is different from the one of sodium. Finally, specific receptors for the regulation of the chloride secretion are located on the basolateral side of the chloride cell membrane. The sensitivity to bicarbonate could imply a Cl/bicarbonate exchange in the basolateral side.

These characteristics are shown in the figure presented below. The intention is to utilize this model for further experimentation in order to explain the known features or discover new ones for the osmoregulatory function of the chloride cells.

Diagram showing the main characteristics of the chloride cell

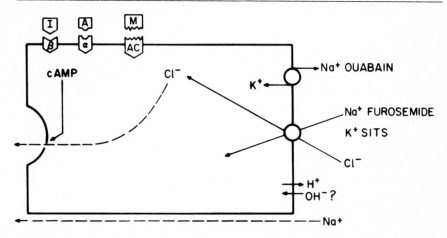

REFERENCES

Bevelander, G. (1935). A comparative study of the branchial epithelium in fishes, with special reference to extrarenal excretion. J. Morphol. 57: 335-352.

Bevelander, G. (1936). Branchial glands in fishes. J. Morphol. 59: 215-224.

Brown, J. and Zadunaisky, J.A. (1982). Fluid movements across anion-secreting epithelia. Fed. Proc. 41: 1266.

Burns, J. and Copeland, D.E. (1950). Chloride excretion in the head region of Fundulus heteroclitus. Biol. Bull. Mar. Biol. Lab., Woods Hole 99: 381-365.

Degnan, K.J., Karnaky, Jr., K.J., and Zadunaisky, J.A. (1977). Active chloride transport in the in vitro opercular skin of a teleost (Fundulus heteroclitus), a gill-like epithelium rich in chloride cells. J. Physiol. 271: 155-191.

Degnan, K.J. and Zadunaisky, J.A. (1979). Open-circuit sodium and chloride fluxes across isolated opercular epithelia from the teleost Fundulus heteroclitus. J. Physiol. 294: 484-495.

Degnan, K.J. and Zadunaisky, J.A. (1980a). Passive sodium movements across the opercular epithelium: the paracellular shunt pathway and ionic conductance. J. Membrane Biol. 55: 175-185.

Degnan, K.J. and Zadunaisky, J.A. (1980b). Ionic contributions to the potential and current across the opercular epithelium. Am. J. Physiol. 238: R231-R-239.

Foskett, J.K., Logsdon, C.D., Turner, T., Machen, T.E., and Bern, H.A. (1981). Differentiation of the chloride extrusion mechanisms during seawater adaptation of the teleost fish, the cichlid Sarotherodon mossambicus. J. Exp. Biol. 93: 209-224.

Foskett, J.K., Machen, T.E., and Bern, H.A. (1982). Chloride secretion and conductance of teleost opercular membrane: Effect of prolactin. Am. J. Physiol. 242: R380-R389.

Henderson, I.W., Chan, D.K.O., Sandor, T. and Chester Jones, I. (1970). The adrenal cortex and osmoregulation in teleosts. Mem. Soc. Endocrinol. 18: 31-55.

Karnaky, Jr., K.J., and Kinter, W.B. (1977). Killifish opercular skin: a flat epithelium with a high density of chloride cells. J. Exp. Zool. 199: 355-364.

Karnaky, Jr., K.J., Degnan, K.J., and Zadunaisky, J.A. (1979). Correlation of chloride cell number and short-circuit current in chloride-secreting epithelia of Fundulus heteroclitus. Bull. Mt. Desert. Isl. Biol. Lab. 19: 109-111.

Keys, A.B. (1931). The heart-gill preparation of the eel and its perfusion for the study of a natural membrane in situ. Z. verol. Physiol. 15: 352-363.

Keys, A.B., and Willmer, E.N. (1932). "Chloride-secreting cells" in the gills of fish with special reference to the common eel. J. Physiol. 76: 368-378.

Krasny, E. and Zadunaisky, J.A. (1978). Ion transport properties of the isolated opercular epithelium of Fundulus grandis. Bull. Mt. Desert Isl. Biol. Lab. 18: 117-118.

Maetz, J. (1969). Seawater teleosts: Evidence for a sodium-potassium exchange in the branchial sodium-excreting pump. Science 166: 613-615.

Maetz, J., and Bornancin, M. (1975). Biochemical and biophysical aspects of salt excretion by chloride cells in teleosts. Fortschr. Zool. 23: 322-362.

Marshall, W.S. (1977). Transepithelial potential and short-circuit current across the isolated skin of Gillichthys mirabilis (Teleostei: Gobiidae), acclimated to 5% and 100% seawater. J. Comp. Physiol. B114: 157-165.

Marshall, W.S., and Bern, H.A. (1979). Teleostean urophysis: urotensin II and ion transport across the isolated skin of a marine teleost. Science 204: 519-521.

Marshall, W.S., and Nishioka, R.S. (1980). Relation of mitochondria-rich chloride cells to active chloride transport in the skin of a marine teleost. J. Exp. Zool. 214: 147-156.

Mayer-Gostan, N., and Zadunaisky, J.A. (1978). Inhibition of chloride secretion by prolactin in the isolated opercular epithelium of _Fundulus heteroclitus_. Bull. Mt. Desert Isl. Biol. Lab 18: 106-117.

Mendelsohn, S.A., Cherksey, B., and Degnan, K.J. (1981). Adrenergic regulation of chloride secretion across the opercular epithelium: The role of cyclic AMP. J. Comp. Physiol. B145: 29-35.

Moreno, J. (1975a). Blockage of gallbladder tight junction cation-selective channels by 2,4,6-triaminopyrimidinium (TAP). J. Gen. Physiol. 66: 97.

Motais, R., Garcia-Romeu, F. and Maetz, J. (1966). Exchange diffusion effect and euryhalinity in teleosts. J. Gen. Physiol. 50: 391-422.

Potts, W.T.W., and Eddy, F.B. (1973). Gill potentials and sodium fluxes in the flounder _Platichthys flesus_. J. Cell. Comp. Physiol. 87: 29-48.

Rowing, G.M., and Zadunaisky, J.A. (1978). Inhibition of chloride transport by acetylcholine in the isolated opercular epithelia of _Fundulus heteroclitus_. Presence of a muscarinic receptor. Bull. Mt. Desert Isl. Biol. Lab. 18: 101-104.

Scheffey, C., Foskett, J.K. and Machen, T.E. (1983). Localization of ionic pathways in the teleost opercular membrane by extracellular recording with a vibrating probe. J. Membrane Biol. 75: 193-203.

Shuttleworth, T.J. (1978). The effect of adrenaline on potentials in the isolated gills of the flounder (_Platichthys flesus_ L.). J. Comp. Physiol. 124: 129-136.

Smith, H.W. (1930). The absorption and excretion of water and salts by marine teleosts. Am. J. Physiol. 93: 485-505.

Ussing, H.H. (1949). The distinction by means of tracers between active transport and diffusion. Acta Physiol. Scand. 19: 43-51.

Ussing, H.H., and Zerahn, K. (1951). Active transport of sodium as the source of electric current in the short-circuited isolated frog skin. Acta Physiol. Scand. 23: 110-127.

Zadunaisky, J.A. (1966). Active transport of chloride in frog cornea. Am. J. Physiol. 211: 506-512.

Zadunaisky, J.A. (1978). Transport in eye epithelia: The cornea and crystalline lens. In: Membrane Transport in Biology, eds. Giebisch, Tosteson, and Ussing. Vol. III, pp. 307-335, Springer-Verlag, Berlin.

Zadunaisky, J.A. (1982). "Chloride Transport in Biological Membranes", Academic Press, New York.

Zadunaisky, J.A. (1984). The chloride cell of the gill and the paracellular pathways. In: Fish Physiology, Vol. 10, D. Randall (ed.), Academic Press, New York, in press.

Supported by NIH Research Grants EY 01340 and GM 25002.

Control of the blood osmolarity in fishes
with references to the functional anatomy of the gut

R.KIRSCH, W.HUMBERT and J.L.RODEAU

I. INTRODUCTION

Fishes have to face various osmotic pressures in their environment ranging from about 1100 mOsmL^{-1} in sea water (SW) to nearly 0 mOsm. L^{-1} in fresh water (FW) or to 3200 mOsm.L^{-1} for species like *Aphanius dispar* living near the dead-sea (Lotan and Skadhauge, 1972). Patterns of osmoregulation in fishes have been extensively reviewed by Evans (1979) and will only be summarized.

Primitive fishes live mainly in SW except for Lampreys and a few elasmobranch species. They have no or little osmoregulation. Hagfishes have blood isoosmotic to SW with essentially inorganic components, chondrichthyans have blood slightly hyperosmotic to SW with inorganic components and high urea accumulation.

Bony fishes, except for the *Crossopterygian*, *Latimeria*, regulate their blood chemistry to maintain a nearly constant osmotic pressure of about 300-400 mOsm.L^{-1} in the same range as all higher Vertebrates. The teleosteans, diversified during the tertiary period in a substantial number of species, are the most submitted to experimental analysis. Stenohaline species, strongly adapted to FW are hyperosmotic regulators. Stenohalines species strongly adapted to SW are hypo-osmotic regulators, and a few euryhaline species remain able to shift from one environment to the other either during their biological cycle or during experimental transfer.

The general patterns of osmoregulation summarized in fig.1 are essentially based on early investigations from Smith (1930, 1932), Keys (1931, 1933), Keys and Willmer (1932), Krogh (1939), followed by

many contributions using radioisotopic fluxes and bioelectrical measurements reviewed by Maetz (1971, 1974, 1976), Skadhauge (1974), Kirschner
(1979), Evans (1979), Groot *et al.*(1983), Lahlou (1970, 1983).

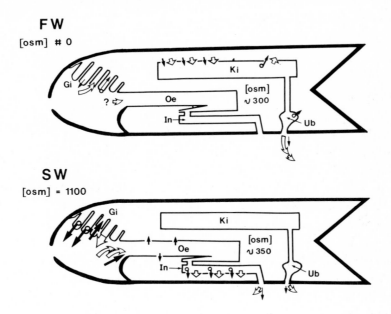

*Fig.1. Diagrams of osmoregulatory patterns in sea water and
fresh water teleosteans : black arrows, ion fluxes;
white arrows, water fluxes; Gi, gills; In, intestine;
Ki, kidney; oe, oesophagus, Ub, urinary bladder.*

The skin is nearly impermeable to water and ions (Fromm, 1968,
Kirsch, 1972) and the gills are the essential site of passive exchanges
owing to their great area (Gray, 1954, Byczkowska-Smyk, 1958) and to
the close contact of the vascular bed and the outer epithelial layer.

In hyperosmotic regulators (FW) the excess of water osmotically
gained through the gills is excreted in urine. Most of the osmoregulatory work consists in reabsorbing monovalent ions from the glomerular
filtrate essentially by the nephron tubule and secondarily by the
urinary bladder (Lahlou, 1970). Ion losses in urine and by diffusion
through the gills are compensated by active ion transport by the chloride cells of the gills. The gut is not an osmoregulatory organ and
drinking will only increase the osmoregulatory work of kidneys. Drinking rate is generally low in FW species except in some euryhaline
species (review in Evans, 1979). For example, premigrating silver eels

drink as much in FW as eels adapted to SW (Kirsch and Mayer-Gostan, 1973).

In hypo-osmotic regulators (SW) the water loss through the gills and in urine can only be replaced by drinking. Urine flow is very low and only carries nitrogen and divalent ions. The gut is the primary organ of osmoregulation in all hypo-osmoregulating vertebrates as well as in hypo-osmoregulating invertebrates like *Artemia salina* (Croghan, 1958 a,b). Ingested SW contains,besides the needed water, an important load of Na^+ and Cl^- ions. They are absorbed in the internal medium where they increase the NaCl load already created by diffusion of Na^+ and Cl^- down their electrochemical gradient through the gills. These ions have to be excreted secondarily by the chloride cells of the gills. Therefore an efficient processing of the drinking water is necessary to limit energy expenditure in the last regulation step through the gills. Ingested SW is processed in the gut in two essential steps : in the oesophagus Cl^- and Na^+ ions are absorbed down their electrochemical potential difference through the epithelium without serosa-to-mucosa water losses (Kirsch *et al.*1975, Hirano and Mayer-Gostan 1976, Kirsch, 1978). In the intestine active ion transport builds up a transepithelial osmotic pressure difference which drives water from lumen to extracellular fluid.

The cellular and subcellular mechanisms of ion absorption in intestine have been recently reviewed in the symposium on "Intestinal transport" by Groot *et al.*(1983), Lahlou (1983), Leray and Florentz (1983). The present paper will summarize general patterns of the gut osmoregulatory function, give further information on eosophageal ions absorption as related to mucus layer and discuss localization of intestinal ions and water absorption on the basis of gross structural and functional observations.

II. GENERAL PATTERNS OF THE GUT OSMOREGULATORY FUNCTION

A.Drinking rate and drinking behaviour
Very little is presently known about the drinking behaviour.

The drinking rate has generally been measured with non-absorbable markers added to the external medium (e.g.polyethylene glycol 14C, colloïdal gold, polyvinyl pyrrolidone 125I) and autopsy of animals after a few hours to determine the gut content. In euryhaline species

drinking rates increase generally with external salinity from FW to SW,
but in higher salinities gut efficiency is increased with constant
drinking rate (Skadhauge, 1976). Very large differences appear in
data concerning SW drinking in different species (Evans, 1979). This
may be related not only to interspecific variability but also to environmental temperature, physiological maturation state of the animal or
experimental choc effects. The SW eels reduce considerably drinking
rate from 5.4 ml.h^{-1}kg^{-1} at 25°C to 0.24 at 5°C(Motais and Isaïa,1972).
The yellow eels behave as typical FW fishes and drink much less than
premigrating silver eels (Gaitskell and Chester Jones, 1971). The
buffalo sculpin increases its drinking rate to normal levels in days
following capture (Sleet and Weber, 1982) and in the eel infusions of
epinephrine decrease the drinking rate (Kirsch and Guinier, 1978). The
large intraspecific variations in drinking rates led to the interpre-
tations of intermittent drinking in sculpins (Foster, 1969).

An other approach to study drinking behaviour consists to
cannulate the oesophagus and to record continuously the drinking rate.
In the eel short experiments of that type (Hirano, 1974) or long-term
experiments (Kirsch and Meister, 1982) demonstrated that drinking
occurs continuously at a very constant rate in the undisturbed animal.
A similar procedure (Sleet and Weber, 1982) led to the conclusion that
the sculpin were "intermittently ingesting SW in small volumes or sip-
ping". A more or less continuous ingestion of SW is important as it
allows a slow progression of the ingested water over the gut epithelium
and successive steps of treatment.

The efficiency of the gut is high and most of the monovalent
ions and water are absorbed while divalent ions remain in the lumen
and are voided with feces (Smith, 1930; review by Evans, 1979). In the
sculpin (Sleet and Weber, 1982) absorptions estimated from the rectal
fluid composition were 97% for Na$^+$, 95% for Cl$^-$ and 69% for water,
which are in the same range then data previously reported. Efficiency
of ion and water absorption is increased in the intestine of euryhaline
fishes in SW compared to FW (Utida *et al.*, 1967; Utida *et al.*, 1969;
Skadhauge, 1969). Functional adaptations to SW processing appear even
in FW before catadromous migration in eel (Utida *et al.*, 1967) and
salmon (Collie and Bern, 1980).

B.Oesophagus

A progressive processing of SW in the gut was obvious
from the *in vivo* observations of the luminal concentrations of monova-

lent ions in the gut of the eel (Sharratt *et al.*, 1964) or the trout
(Shehadeh and Gordon, 1969). These authors reported low concentrations
of NaCl in the stomach (eel) and intestine (eel and trout) in reference to ingested SW. This led to the discovery of the osmoregulatory
function of the oesophagus demonstrated in the eel by perfusion experiments *in vivo* (Kirsch and Laurent, 1975; Kirsch *et al.*, 1975) and by
isolated sac technique (Hirano and Mayer-Gostan, 1976). Other FW and
SW teleostean species were analysed with the isolated sac technique
(Kirsch, 1978) to confirm previous results. Recently Sleet and Weber
(1982) assessed the oesophageal osmoregulatory function in the sculpin
by *in vivo* perfusion experiments. The oesophagus of FW animals is impermeable to ions and water and very low serosa-to-mucosa net fluxes
may be related to mucus secretion. The oesophagus of SW animals is very
permeable to Na^+ and Cl^-. This allows fast absorption down the electrochemical gradient between lumen and serosa of about 50 to 70% of the
ions ingested with SW. The oesophagus remains impermeable to water
and no noticeable net fluxes occur down the osmotic gradient between
serosa and lumen. The functional adaptation to SW is substantiated by
important structural modifications in the eel (Laurent and Kirsch, 1975;
Yamamoto and Hirano, 1978); these modifications fit very well with the
comparative anatomy of different FW and SW species (Meister *et al.*,
1983). In summary, FW species have a thick stratified epithelium with
very numerous mucous cells covering a poorly vascularized conjonctive
layer; SW species have the same structure in the beginning of the oesophagus but the stratified epithelium is progressively replaced by a
simple microvillous columnar epithelium with often dilated intracellular spaces. Important foldings increase the epithelial area in SW and
the underlaying vascular bed becomes very dense with large blood vessels.
The mucous layer is supposed to wip preys from excess of water during
feeding. The microvillous epithelium is probably the low resistance
pathway for rapid monovalent ion diffusion in SW but does not explain
water impermeability. Recently, determinations by autopsy of normal
animals, of chloride concentrations in the luminal fluid at different
levels of the gut substantiated the generality of osmoregulatory function in SW teleosteans. The concentrations of Cl^- at the end of the
oesophagus was shown to be 36-67% of SW values, depending on the species
(Kirsch and Meister, 1982).

C.Stomach

The stomach has little role, if any, to play in osmoregulation. In isolated stomachal sacs from eels filled with SW, Hirano and

Mayer-Gostan (1976) observed a Cl⁻ net flux from serosa-to-mucosa and an opposite net water flux diluting luminal fluid. However, no significant Cl⁻ or water net fluxes could be observed during *in vivo* perfusions of the stomach in starved eels (Kirsch and Meister, 1982). An osmotic dilution of gastric content was also reported during *in vivo* perfusion of the stomach in the cod (Holstein, 1979b) but the perfusate was pure SW which does not represent a physiological stomacal content (Kirsch and Meister, 1982).

D. Intestine

The intestine plays the most active part of gut osmoregulation and its function was clearly demonstrated by Skadhauge (1969-1974). Eel intestines were perfused *in vivo* with a saline solution corresponding to half diluted SW. Recycling of the perfusate allowed determination of chloride, sodium , and water net fluxes as a function of the progressive modifications of the perfusate in good physiological conditions. Early dilution of the perfusate occured by a serosal-to-mucosal water net flux along the osmotic gradient together with active ion absorption from mucosa-to-serosa against the electrochemical gradient. During further dilutions, the water flux decreased progressively to zero and reversed to an increasing net water flux from mucosa-to-serosa. The osmotic concentrations in lumen when net water flux reversed ("turning-point") was higher than in plasma, indicating a water absorption against the over-all osmotic gradient. Moreover, the turning-point osmolality differences and the water net fluxes appeared directly proportional to the intensity of transmural flow of NaCl in different experiments. These findings were in agreement with the Diamond's model (1964) for fluid absorption through epithelia, the uphill water absorption being linked to local osmosis due to ion transport from lumen to cell and cell to intercellular spaces. This model is substantiated by the detailed structural and functional analysis of the intestinal epithelium of the winter flounder (Field *et al.*, 1978). The epithelium is composed of parallel,long (60 μm) and narrow (3·5 μm),cells associated by typical junctional complexes at their apical poles. There is little distension of the apical part of the intercellular spaces (12μm) but deeper lateral parts appear dilated with restrictions at the level of numerous desmosomes. The lateral surface is amplified by infoldings which communicate with lamellar structures within the cell. Distension of lateral spaces is particularly evident in epithelia stripped from the muscular layer and used in Ussing chambers. An important compartment of great area exists between enterocytes for water absorption by local osmosis. The pattern of water absorption established by Skadhauge is only limited by the fact that, in all teleosts investigated in SW, the luminal content is more diluted than plasma from the beginning of intestine (Sharratt *et al.*, 1964; Shehadeh and Gordon, 1969; Kirsch and Meister, 1982). Correlatively the first step of luminal osmotic dilution may only exist in the immediate post-pyloric section of the intestine. In the extensively studied eel, no osmotic dilution of luminal fluid normally occurs as the prepyloric 158 mMol $Cl^-.L^{-1}$ (Kirsch and Meister, 1982) are already very low as compared to the 210 mMol $Cl^-.L^{-1}$ corresponding to the experimental mean turning-point for 0 net water fluxes (Skadhauge, 1974).

The mechanisms of ion transport, which provide the driving force for water absorption through the intestinal epithelium, were extensively submitted to experimental analysis with *in vitro* techniques (review by Groot *et al.*, 1983; review by Lahlou, 1983; Badia and Lorenzo,1982; Loretz, 1983). On the basis of serosa-negative potential in reference to mucosa, an electrogenic chloride pump was postulated as driving force by many authors, Na^+ following the electrochemical gradient. An active Cl^- pump was particularly substantiated in the eel intestine (Ando *et al.*,1975; Ando, 1975, 1980, 1981; Ando and Kobayashi, 1978). Field *et al.*(1978) suggested an other model in three steps : Na^+ and Cl^- are absorbed in the ratio 1/1 by the brush-border, Na^+ is pumped from cell to the lateral spaces and Cl^- follows passively the electrochemical gradient, Na^+ diffuses partly back to the lumen through the apical junction permeable to Na but very little permeable to Cl^- which moves downhill to the serosal side. The cation selectivity of the apical junction is the basis of the serosa-negative transepithelial potential. This model is well substantiated by intracellular recordings of Cl^- activity and electrical potential differences across the mucosal membrane of the enterocyte (Duffey *et al.*, 1979). Cl^- is accumulated in the cell against an important electrochemical potential difference only if Na^+ is absorbed in the cell down its electrochemical potential difference and energizes coupled Cl^- absorption.

E.Regulation of water intake

The drinking rate is regulated by complex mechanisms including general endocrine control of the hydromineral balance (review by Holmes and Pearce, 1979) and control of the intestinal transport efficiency particularly by ACTH and cortisol (Oide and Utida, 1967, 1968) and urotensin II (Mainoya and Bern, 1982) for SW adaptation and

by Prolactin (review by Hirano *et al.*, 1976, Hirano, 1980a,b) and urotensin I (Mainoya and Bern, 1982) for FW adaptation. The drinking rate
is also more specifically controled : drinking is induced by angiotensin II (Hirano *et al.*, 1978; Holstein and Brigel, 1981) and by the
brain-stem of the central nervous system in response to cellular or
extracellular dehydration (Hirano, 1974). In opposition extracellular
volume expansion by perfusion inhibits drinking in SW (Hirano, 1974;
Holstein and Brigel, 1981). Distention of the stomach or intestine in
the SW eel inhibits drinking (Hirano, 1974); the same result is obtained in the cod with intestinal perfusion (Holstein, 1979a) but drinking
is restored by acidification of the intestine. The nature of the external medium is also a stimulus for drinking rate regulation, Cl^- ions
at higher concentration than 20 $mMol.L^{-1}$ specifically induce drinking
behaviour whereas FW inhibits drinking (Hirano, 1974).

A combination of these regulations may now explain for example
the very complex biphasic pattern of drinking rate in function of time
during the adaptation of the FW eel to SW (Kirsch and Mayer-Gostan,
1973). Immediately after the FW to SW transfer the eel drinks a lot
as specific reaction to high Cl^- concentration in the environment,
the resulting expansion of the stomach and intestine reduces drinking
for several hours. This inhibition is overcome on the following days
by stimulation linked to dehydration of the animal and drinking increases to reach a maximal value between 4-7 days SW adaptation. Then,
the intestine becomes fully adapted to SW processing under cortisol
stimulation and probably distension inhibition completely disappears.
Finaly in two or three weeks, the drinking rate decreases to steady-
state levels with restoration of the hydro-mineral balance.

III. MUCUS AND OESOPHAGEAL ION ABSORPTION

The mucus coating has been assessed for a long time to protect
the skin for adaptation to high salinity environments (Portier and
Duval, 1922) although this is not clearly established yet (Marshall,
1978; Shephard, 1981). Mucous cells are also largely distributed in
all parts of the gut and in fishes are particularly abundant in the
oesophageal epithelium (Meister *et al.*, 1983). The mucus coating in
the gut has been recently demonstrated to constitute an apical compartment of functional importance over the epithelial cells, modifying
diffusion kinetics of Na^+ ions and organic components (for a review
see Gilles-Baillien, 1983). Particularly demonstrative are the functions

of mucus in maintaining a neutral pH at the cell contact against a pH
of 2 in the lumen of the duodenum (Flemström, 1983) or the role of the
thick colonic mucus layer in maintaining an acid microclimate at cell
contact which favour short-chain fatty acid absorption (Sakata and
Von Engelhardt, 1981a,b; Von Engelhardt and Rechkemmer, 1983). In the
oesophageal mucus layer of the SW teleost, *Rhombosolea retiaria*,
Shepart (1982) using ion selective microelectrodes, reported gradients
of Na^+, K^+ and Ca^{++} activities from apical SW to cell contact.

An important diffusion barrier for Cl^- ions in mucus would be
of interest to explain why, in the SW eel, the Cl^- concentrations in
the luminal fluid decrease from 520 to about 200 $mMol.L^{-1}$ at the beginning of the oesophagus and remain nearly constant down the whole organ.
A first explicative hypothesis was that highly specialized absorbing
area may be present at the beginning of oesophagus to decrease Cl^- concentration in ingested SW; this was not substantiated by structural
analysis (Meister *et al.*, 1983). A second hypothesis was that thick
unstirred water compartments in mucus over the oesophageal epithelium
may maintain standing gradients of Cl^- concentration between SW in
lumen and the cell surfaces, the bulk Cl^- concentration in these compartments being nearly constant. Recently, thick mucus layers with
different structures were observed covering the oesophageal epithelium
in the eel (Humbert *et al.*, 1983). Mucus is maintained for observation
by rapid deep-freezing of fresh tissue samples followed by freeze-drying in dry ice. The mucus appears in the beginning of oesophagus as a
dense layer anchored firmly to the microridges of epithelial cells.
This dense mucus is progressively replaced by a fibrillar mucus (fig.2)
covering microvillous columnar cells. X-ray microprobe analysis of Cl^-
was performed on fractures of freeze-dried samples at three levels of
the mucus fibers shown in fig.2 : lumen contact, middle part and cell
contact (table 1). In reference to blood plasma values it appears that
92% of the lumen to blood Cl^- gradient is supported in the mucus layer.

TABLE 1. Chloride X-Ray microanalysis in fibrous oesophageal mucus
from the sea water eel : x = counts per 20 seconds

		n	$\bar{x}$	s.e.m	p
Mucus fibres	lumen contact	10	6077	185	<0.001
	middle part	10	4693	164	<0.01
	cell contact	10	3565	299	n.s.
Plasma		14	3338	93	

lumen

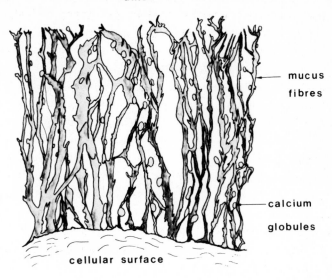

mucus
fibres

calcium

globules

cellular surface

Fig.2. Drawing from an original micrograph of the fibrous
mucus covering the microvillous oesophageal epithelium
of the sea water eel : thickness of the mucus layer, 50μm.

The previous technique, however, was relatively coarse, due
to the fact that ions were precipitated as salts from their original
solution to neighbouring mucus fibers during freeze-drying.

A complementary analysis was thus performed with Cl^- selective
microelectrodes (Cl^- ion exchanger, Corning 477/315) following exactly
the technique described by Shepard for cation activity analysis in
mucus. A patch of oesophagus dissected from its muscular layers was
placed between two perpex chambers and perfused with an artificial
extracellular solution (AECS) on the serosal side and with artificial
SW (ASW) or artificial FW (AFW) or AECS on the mucosal side. Great
care was taken to avoid any contact with the mucus layer during preparation of the sample. Preliminary impalements with KCl filled conventional potential electrodes showed no significant potential difference
across the mucus layer. Calibration of the Cl^- selective electrodes
with sodium chloride solutions gave a slope of 55 mV per 10 fold increase in activity (temperature 28°C). The Cl^- activities in mucus and
in SW were calculated from the Cl^- selective electrode potential deflec-
tions (ΔE) and the theoretical Cl^- activity value in the AECS (96 mMol.
L^{-1}). The Cl^- activity in FW (1,4 mMol.$Cl^-.L^{-1}$) was the theoretical one,

the electrode response being not linear in such diluted solution.
Recordings of ΔE are shown in fig.3, 4 and 5.During progressive impalements, sometimes, the tip of the microelectrode, fortunately, made his
way in the mucus without sticking to the mucous fibers, then recordings
showed a progressive increase in ΔE to a plateau characteristic of the
microelectrode coming into contact with the cell surface (first parts
of traces in fig.1, posterior oesophagus 1 and anterior oesophagus 1).
Usually, sticking occured and the mucus was draged by the electrode;
this resulted in coarse penetration to more or less deep parts of the
mucus layer (initial part of traces in fig.1, posterior oesophagus 2;
fig.2 and 3). In the eel, the Cl^- gradient in mucus perfused with SW,
thus is less clearly recorded than the cations gradients reported by
Shepard (1982), owing to the thickness of the oesophageal mucus which
was 40-150 µm in the eel and 50 to 1200 µm in *Rhombosolea* (observations
reported refer respectively to 50-100 and 800 µm impalements).

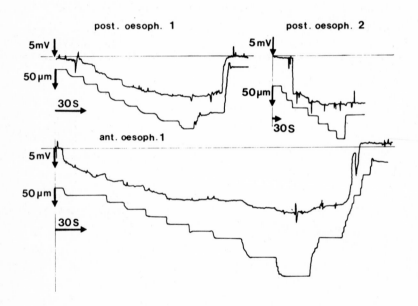

Fig.3. *Drawings of original recordings of oesophageal mucus*
impalement with Cl^- selective microelectrodes (mucus
perfused with artificial sea water). Upper traces : Cl^-
selective microelectrode potential; lower traces :
position of the tip of the microelectrode.

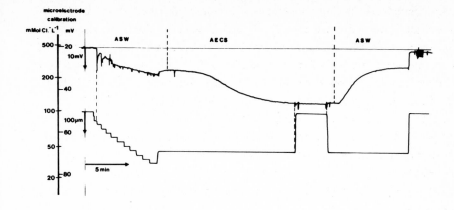

Fig.4. *Drawings of original recordings of oesophageal mucus*
impalement with a Cl⁻ selective microelectrode. Mucus
perfused alternatively with artificial sea water (ASW)
and artificial extracellular solution (AECS) used also
on serosal face of the epithelium. Other captions as in
fig.3.

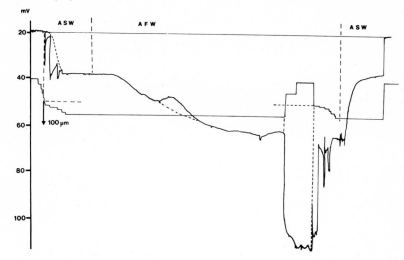

Fig.5. *Drawings of original recordings of oesophageal mucus*
impalement with a Cl⁻ selective microelectrode. Mucus
perfused alternatively, with ASW and artificial fresh
water (AFW). Other captions as in fig.3 and 4.

Nevertheless, Cl$^-$ potentials, in SW and in mucus at the cell contact, were very stable and reproducible. The Cl$^-$ potential differences across mucus could be measured in 4 animals on different patches of anterior and posterior oesophagus. They allow calculation of the part of the Cl$^-$ activity gradient between lumen and serosa supported in mucus (Table 2). No significant difference appears between mucus over anterior and posterior oesophagus. The mucus layer supports from 52 to 82% of the over-all transoesophageal Cl$^-$ gradient in different animals. Modifications of the Cl$^-$activity gradient in mucus were obtained by replacement of initial SW on mucosal side by the same solution (AECS) as on the serosal side (fig.4) or by FW (fig.5). With AECS on both sides of the preparation (fig.4) the Cl$^-$ gradient disappears completely and no Cl$^-$ potential difference can be observed when the Cl$^-$ electrode is pulled out of the mucus layer (Δ E in reference to SW = 25.3 mV, s.d. = 0.6, n = 9). The gradient is exactly restored when SW is perfused again on the mucosal side. This is a good indication that the gradient may be purely diffusive and that no Cl$^-$ to mucus binding occurs as described for cations in the amphibians skin mucus (Kirschner, 1978).

TABLE 2 : Chloride selective microelectrode potential differences and chloride activity across oesophageal mucus from the sea water (SW) eel : $\overline{mV}$, mean potential difference between SW and the mucus at cell contact; $\overline{a}$, mean Cl$^-$ activity in mucus at cell contact (activity SW : 277 mMolCl$^-$.L^{-1}); %, percentage of the lumen-to-serosa Cl$^-$ activity gradient supported in the mucus layer

	Anterior oesophagus					Posterior oesophagus				
	$\Delta \overline{mV}$	sd	n	$\overline{a}$	%	$\Delta \overline{mV}$	sd	n	$\overline{a}$	%
A$_1$	14.5	3.9	6	151	70	9.9	2.5	8	183	52
A$_2$	18.2	0.6	4	129	82	12.3	2.6	6	165	62
A$_3$	16.5	2.3	11	138	77	15.6	1.8	9	144	73
A$_4$	-	-	-	-	-	18.4	2.6	14	128	82

With replacement of SW by FW the Cl⁻ activity gradient reverses (fig.5) and can be restored to initial values with replacement of FW by SW. Cl⁻ potentials in reference to SW are 101 mV in FW (s.d. = 5.8, n=10) and 40 mV in the mucus at cell contact (s.d. = 4.9, n=10), i.e. Cl⁻ activities of 1.4 mMol in FW, 52 mMol in mucus and 96 mMol in serosal solution. The mucus supports 53% of the over-all Cl⁻ gradient in FW. A possible correlation between Cl⁻ potential differences and thickness of the mucus layer was also checked (fig.6) but no correlation appeared.

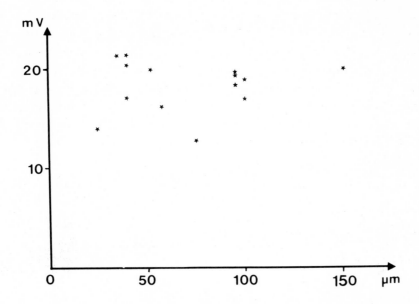

Fig.6. *Cl⁻ selective electrode potential differences (mV) across the oesophageal mucus layer as a function of its thickness (μm)*

In all experimental conditions the mucus constitutes a diffusion barrier supporting the main part of the over-all luminal-to-serosa Cl⁻ gradient. The functional importance of this diffusion barrier for SW processing has to be explained,only hypotheses can be put forward presently. From the structure of the fibrous mucus and its property to dilute progressively in luminal water, it is likely that physical resistance to longitudinal water flow may increase in the mucus layer from lumen to cell contact. If, as expected from structural observations, the thickness of the mucus layer decreases from the beginning to the end of the oesophagus, then SW could flow down the oesophagus through the mucus layer with different speeds and allows a progressive dilution as illustrated in fig.7. As stated by Shepard (1982), the mucus must not only support a component of the over-all ionic gradient but also of the osmotic gradient. From preceeding observations *in situ* (Cl⁻ X-Ray microanalysis) the ion concentration at the cell contact would be hypertonic to serosal fluid by only 8% which would explain why, even with a very leaky oesophageal epithelium, no net water flux from serosa-to-mucosa is observed *in vivo* whereas water permeability is increased *in vitro* when mucus is wiped off from the oesophageal epithelium (Kirsch, 1978). The proportions of Cl⁻ gradient in mucus to over-all gradient are higher for *in situ* observations (92%) than for *in vitro* observations (52-82%) in the eel and also much higher than for Na⁺ gradients (9,1-16,6%) reported by Shepard in *Rhombosolea*.The discrepancy between eel and *Rhombosolea* can not presently be explained as nothing is known about the structure and the localization of mucus used in the last species. In the eel discrepancy between *in vitro* and *in vivo* observations can easily be accounted for by the absence of tissular drainage by the blood flow which *in vitro* may add in the system a serosal gradient to the overall gradient from lumen to serosal fluid.

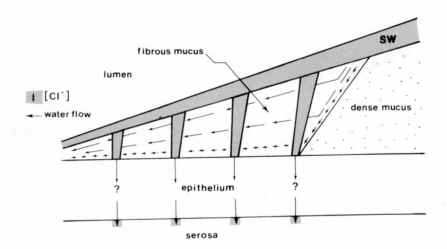

Fig.7. Hypothetical model of water flow and Cl⁻diffusion in the oesophageal mucus : arrow in fibrous mucus are proportional to hypothetical water flows, the breadth of dotted bars is proportional to the Cl⁻ activity at different levels.

Nothing is known about the properties of the dense mucus covering the anterior oesophagus. It may only be mentioned that, on some oesophageal patches, it was impossible to get impalements of the mucus layer, which sticked to the tip of the electrode and was completely forced back without impalement.

The functional hypothesis concerning mucus has now to be checked by more detailed X-Ray analysis on complete longitudinal sections of oesophagus and its mucus layer.

Last, it must be emphazised that diffusive pattern of oesophageal ion absorption presently described has to be completed by the possibility of active ion transport systems substantiated by structural analysis of the microvillous epithelium (Meister *et al.*,1983) and Na^+ active transport reported in the flounder oesophagus (Parmelee and Renfro, 1981).

IV. LOCALIZATION OF IONS AND WATER ABSORPTION IN THE INTESTINE

Investigations on the mechanisms of ion transport in the fish intestine are usually performed on posterior or middle intestinal patches, which are easier to stripp from the muscular layers and to stretch in Ussing Chambers. Some comparisons of transporting efficiency were done between posterior- and anterior intestine in the eel (Ando *et al.*, 1975) or between posterior- and middle intestine in the goby (Loretz, 1983); the posterior parts of the intestine appeared to be the most potent in active ion transport. But in the trout the anterior intestine was the only part transporting Na^+ *in vitro* (Bensahla-Talet *et al*, 1974). *In vivo* the anterior intestine of the eel was also more efficient in Cl^- and water absorption than the posterior intestine during perfusion experiments. Moreover, low transport efficiencies in the anterior intestine are unlikely, *in vivo*, as the Cl^- concentration in luminal fluid decreases very steeply, from the stomach to the anterior intestine in the 9 different species analysed (Kirsch and Meister, 1982). Diffusion barriers on serosal side were clearly demonstrated to impair transporting efficiency in intestine by Ando and Kobayashi (1978) with stripped and unstripped intestines. The serosal diffusion barriers are probably greater in the anterior intestine than in the posterior intestine and may explain the low efficiency reported for the anterior intestine *in vitro*. In the ante-

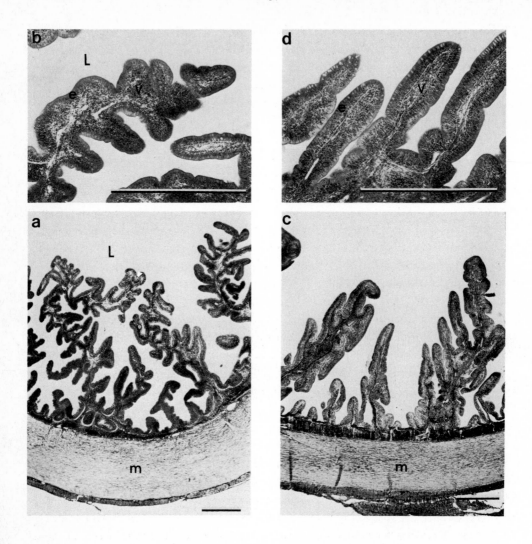

Fig.8. Cross-sections of the anterior intestine from sea
water (SW) and fresh water (FW) eels. Tissue fixation
with Bouin's fluid; stained with Trioxyhematein according to Gabe; a,b, SW eel; c, d, FW eel; e, enterocytes;
l, lumen; m, muscular layer; v, villous core; bars
represent 500 µm.

rior intestine of the starved european eel (fig.8) the intestinal villi considerably increase their surface by infoldings during FW to SW adaptation and the serosal space becomes very narrow. This increases the apical to basal diffusional resistance if blood flow is suppressed. The same modifications linked to SW adaptation were reported for the trout (Macleod, 1978) and an hyperplasia of intestinal epithelium was reported in the american eel (Mackay and Janicki, 1979). In the posterior intestine of the starved european eel (fig.9), villi are small compared to those of anterior intestine and they are less modified during FW to SW adaptation. They are also easily streched in Ussing Chambers (Field *et al.*, 1978). Consequently, serosal diffusion barrier in villi of the posterior intestine is low, *in vitro*, compared to that of anterior intestine. A very important increase in epithelial area was also shown in the pyloric canal of the eel during FW to SW adaptation (Meister, 1982).

Gross structural adaptations substantiate an efficient transport of ions and water in the anterior intestine. Moreover, the structure of the high villi in the anterior intestine could be the support of baso-apical gradients of osmotic pressure in the serosal core, which would allow a very fast water absorption. The existence of such gradients was demonstrated by Jodal *et al.* (1978) in the core of the intestinal villi in the cat, the osmotic gradient ranging from 1200 $mOsm.L^{-1}$ at the apical part to 300 $mOsm.L^{-1}$ at the basis of the villi. This gradient was built up by Na^+ transport and was of essential importance for net water transport (Hallbäck *et al.*1979a,b, 1980). Importance of the villi in ion transport appeared also in *Amphiuma* (Gunter-Smith and White, 1979), where Na^+ dependent solute transport and the basal electrogenic ion transport processes were linked to intestinal villous tissue and not to intervillous epithelium. Nothing is known in fishes about a possible role of serosal compartments in villi. It would be of the greatest interest to achieve detailed morphometric analysis of the villous epithelium and of its vascularization together with analysis of the ion content in the villous core with different experimental luminal fluids.

An important processing of luminal water in the anterior part of the intestine could be functionaly very important as Holstein (1979 a,b) demonstrated, in the cod, that an *in vivo* perfusion of the intestine with SW reduced considerably HCl secretion by the stomach, whereas perfusion with diluted SW allowed normal secretion. A high salinity in the intestine may consequently impair digestive function.

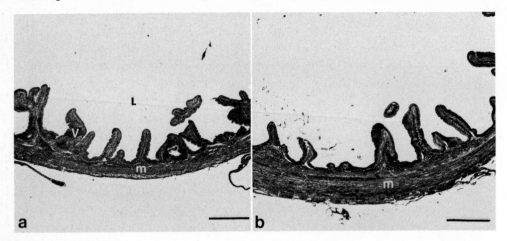

Fig.9. Cross-sections of the posterior intestine from sea-water (a) and fresh-water (b) eels. Captions as in fig.8.

V. CONCLUSIONS

The gut of SW fishes is very efficient in compensating for osmotic water losses through the gills. In first oesophageal step, probably mainly diffusive, a large amount of Cl^- and Na^+ ions are absorbed down their electrochemical gradient without water loss from serosa-to-mucosa. In the second intestinal step, only about one third of the ingested SW salt content has to be transported uphill to carry all the water. This sequential processing certainly saves energy in reference to the early exclusively intestinal model.

The intestinal epithelium has a very simple structure and has no Lieberkühn crypts (Andrew and Hickman, 1974). Moreover, the fishes are poïkilotherm animals and can fast for long periods which enables to separate experimentaly digestive and osmoregulatory processes. Their tissues may also be studied at different temperatures. Some of them are euryhaline and can be adapted to different salinities. All these characteristics made teleosts a very interesting biological material and the intestinal epithelium was intensively studied as a model of ion transporting epithelia. Presently the transepithelial ion transporting mechanisms driving water absorption are well documented and the isolated intestine from teleost has brought large contributions to progress in cellular physiology.

In contrast, very little is known about the compartments preceeding and following the intestinal cells and many complementary investigations are needed.

The mechanisms involved in the Na^+ and Cl^- absorption through the oesophageal epithelium have to be checked *in vitro*, with and without the mucus layer, the system epithelium-plus-mucus being very original by its high monovalent ion net fluxes not coupled to net water fluxes.

The mucus may also play an important role in osmoregulation in the intestine. In the posterior intestine, it could constitute the "Na^+ apical reservoir" stated by Brot-Laroche and Alvarado (1983) to explain sugar and amino-acid transport without Na^+ in the luminal fluid, but it could also be useful to explain the water absorption through the intestine (House and Green, 1965) and the Cl^- absorption through the apical membrane of the enterocyte (Duffey *et al.*, 1979), which continue for a long time after Na^+ has been suppressed in the luminal fluid.

The role of the different parts of the intestine have to be studied, particularly that of the anterior intestine where important structural adaptations are linked to SW adaptation and are not presently integrated in the functional scheme of osmoregulation. The serosal core in villi could possibly have an important part to play in water absorption.

In summary, investigations in cellular physiology have to be completed by analysis focussing on supra-cellular and sub-cellular compartments.

REFERENCES

Ando M. (1975) Intestinal water transport and chloride pump in relation to sea-water adaptation of the eel, *Anguilla japonica*. Comp. Biochem. Physiol. 52A : 229-233.

Ando M. (1980). Chloride-dependent sodium and water transport in the sea water eel intestine. J. Comp. Physiol. 138 : 87-91.

Ando M. (1981). Potassium-dependent chloride transport in the sea-water eel intestine. J. Physiol. Soc. Japan 43 : 282.

Ando M., Utida S., Nagahama H. (1975). Active transport of chloride in eel intestine with special reference to sea water adaptation. Comp. Biochem. Physiol. 51A : 27-32.

Ando M., Kobayashi M. (1978). Effects of stripping of the outer layers of the eel intestine on salt and water transport. Comp. Biochem. Physiol. 61A : 497-501.

Andrew W., Hickman C.P. (1974). In "Histology of the vertebrates" (Mosby C.V. ed.) St.Louis.

Badia P., Lorenzo A. (1982). Preliminary studies on transmural potential and intensity of the short-circuit current in intestine of *Gobius maderensis*. Rev. esp. Fisiol. 36 : 221-226.

Bensahla-Talet A., Porthé-Nibelle J., Lahlou B. (1974). Le transport de l'eau et du sodium par l'intestin isolé de la truite *Salmo irideus* au cours de l'adaptation à l'eau de mer. C.R.Acad.Sc.Paris 278 : 2541-2544.

Brot-Laroche E., Alvarado F. (1983). Mechanisms of sugar transport across the intestinal brush border membrane. In "Intestinal transport" (Gilles-Baillien M. and Gilles R. eds.) Springer Verlag. Berlin. p.147-169.

Byczkowska-Smyk W. (1958). The respiratory surface of the gills in teleosts. Part II. The respiratory surface of the gills in the eel (*Anguilla anguilla L.*), the loach (*Misgurnus fossilis L*) and the perch-pike (*Lucioperca lucioperca* L.). Acta Biologica Cracoviensia 1 : 83-87.

Collie N.L., Bern H.A. (1980). Variations in water transport across the coho salmon posterior intestine during smoltification. Amer. Zool. 20(4) : 873 .

Croghan P.C. (1958a). The mechanism of osmotic regulation in *Artemia salina* L. : The physiology of the branchiae. J. Exp.Biol. 35 : 234-242.

Croghan P.C. (1958b). The mechanism of osmotic regulation in *Artemia salina* L. : The physiology of the gut. J. Exp. Biol. 35 : 243-249.

Diamond J.M. (1964). The mechanism of isotonic water transport. J. Gen. Physiol. 48 : 15-42.

Duffey M.E., Thompson S.M., Frizzell R.A., Schultz S.G. (1979). Intracellular chloride activities and active chloride absorption in the intestinal epithelium of the winter flounder. J.Membrane Biol. 50 : 331-341.

Engelhardt W.v., Rechkemmer G. (1983). Colonic transport of the short-chain fatty acids and the importance of the microclimate. In Falk-Symposium n°36 "Intestinal absorption and secretion". Abstracts p.6.

Evans D. (1979). In "Comparative physiology of osmoregulation in animals" (Maloiy G.M.O. ed.) 1 : 306-390.

Field M., Karnaky Jr. K.J., Smith P.L., Bolton J.E., Kinter W.B.(1978). Ion transport across the isolated intestinal mucosa of the winter flounder, *Pseudopleuronectes americanus*. I. Functional and structural properties of cellular and paracellular pathways for Na and Cl. J. Membrane Biol. 41 : 265-293.

Flemström G. (1983). HCO_3^- secretion by the duodenum. In Falk-Symposium n°36 "Intestinal absorption and secretion". Abstracts p.20.

Foster M.A. (1969). Ionic and osmotic regulation in three species of Cottus (Cottidae, teleost). Comp. Biochem. Physiol. 30 : 751-759.

Fromm P.O. (1968). Some quantitative aspects of ion regulation in teleosts. Comp. Biochem. Physiol. 27 : 865-869.

Gaitskell R.E., Chester Jones I. (1971). Drinking and urine production in the european eel (*Anguilla anguilla* L.). Gen. Comp. Endocr. 16 : 478-483.

Gilles-Baillien M. (1983). Several compartments involved in intestinal transport. In "Intestinal transport" (Gilles-Baillien M. and Gilles R. eds.) Springer Verlag. Berlin. p.103-119.

Gray J.E. (1954). Comparative study of the gill area of marine fishes. Biol. Bull. 107(2) : 219-225.

Groot, J.A., Albus H., Bakker R., Heukelom J.S.v., Zuidema T. (1983). Electrical phenomena in fish intestine. In "Intestinal transport" (Gilles-Baillien M. and Gilles R. eds) Springer Verlag Berlin p. 321-340.

Gunter-Smith P.J., White J.F. (1979). Contribution of villus and intervillus epithelium to intestinal transmural potential difference and response to theophylline and sugar. Biochem. biophys. Acta 557 : 425-435.

Hallbäck D.A., Jodal M., Lundgren O. (1979a). Importance of sodium and glucose for the establishment of a villous tissue hyperosmolality by the intestinal counter current multiplier. Acta Physiol. Scand. 107 : 89-96.

Hallbäck D.A., Jodal M., Sjöqvist A., Lundgren O. (1979b). Villous tissue osmolality and intestinal transport of water and electrolytes. Acta Physiol. Scand. 107 : 115-126.

Hallbäck D.A., Jodal M., Lundgren O. (1980). Villous tissue osmolality water and electrolyte transport in the cat small intestine at varying luminal osmolalities. Acta Physiol. Scand. 110 : 95-100.

Hirano T. (1974). Some factors regulating water intake by the eel, Anguilla japonica. J.Exp. Biol. 61 : 737-747.

Hirano T. (1980a). Prolactin and osmoregulation. J. Endocr. 186-189.

Hirano T. (1980b). Effects of cortisol and prolactin on ion permeability of the eel oesophagus. In "Epithelial transport in the lower vertebrates" (Lahlou B. ed.) p.143-149.

Hirano T. , Mayer-Gostan N. (1976). Eel oesophagus as an osmoregulatory organ. Proc.Nat. Acad. Sci. USA. 73(4) : 1348-1350.

Hirano T., Morisawa M., Ando M., Utida S. (1976). Adaptive changes in ion and water transport mechanism in the eel intestine. In "Intestinal ion transport" (Robinson J.W.L. ed.) MTP Press. Lancaster. p.301-317.

Hirano T., Takei Y., Kobayashi H.(1978). Angiotensin and drinking in the eel and the frog. In "Osmotic and volume regulation". Alfred Benzon Symposium XI, Munksgaard. p. 123-134.

Holmes W.N., Pearce R.B. (1979). Hormones and osmoregulation in the vertebrates. In "Mechanisms of osmoregulation in animals" (Gilles R. ed.) John Wiley and Sons. Chichester. p. 413-533.

Holstein B. (1979a). Gastric acid secretion and water balance in the marine teleost Gadus morhua. Acta Physiol. Scand. 105 : 93-107.

Holstein B. (1979b). Gastric acid secretion and drinking in the Atlantic cod (Gadus morhua) during acidic or hyperosmotic perfusion of the intestine. Acta Physiol. Scand. 106 : 257-265.

Holstein B., Brigel B. (1981). Effects of exogenous angiotensin II in the Atlantic cod , Gadus morhua. Acta Physiol. Scand. 113 : 363-369.

House C.R., Green K. (1965). Ion water transport in intestine of Cottus scorpius. J. Exp. Biol. 42(1) 177-189.

Humbert W. , Kirsch R., Meister M.F. S.E.M. study of the oesophagial mucus layer in *Anguilla anguilla*. J. Fish. Biol., in press.

Jodal M., Hallbäck D.A., Lundgreen O. (1978). Tissue osmolality in intestinal villi during luminal perfusion with isotonic electrolyte solutions. Acta Physiol. Scand. 102 : 94-107.

Keys A.B. (1931). Chloride and water secretion and absorption by the gills of the eel. Z. Vergl. Physiol. 15 : 364-388.

Keys A.B.(1933). The mechanism of adaptation to varying salinity in the common eel and the general problem of osmotic regulation in Fishes. Proc. Roy. Soc. London Ser.B. 112 : 184-199.

Keys A.B. , Willmer E.N. (1932). "Chloride secreting cells" in the gills of fishes with special reference to the common eel. J.Physiol. London 76 : 368-378.

Kirsch R. (1972). The kinetics of peripheral exchanges of water and electrolytes in the silver eel (*Anguilla anguilla* L.) in fresh water and in sea water. J. Exp. Biol. 57 : 489-512.

Kirsch R. (1978). Role of the oesophagus in osmoregulation in teleost fishes. In "Osmotic and volume regulation". Alfred Benzon Symposium XI. Munksgaard. p.138-154. Academic Press, New York.

Kirsch R., Mayer-Gostan N. (1973). Kinetics of water and chloride exchanges during adaptation of the european eel to sea water. J. Exp. Biol. 58 : 105-121.

Kirsch R., Laurent P. (1975). L'oesophage, organe effecteur de l'osmo-régulation chez un téléostéen euryhalin, l'anguille (*Anguilla anguilla* L.) C.R. Acad. Sci. Paris 280 : 2013-2015.

Kirsch R., Guinier D., Meens R. (1975). L'équilibre hydrique de l'anguille européenne (*Anguilla anguilla* L.). Etude du rôle de l'oesophage dans l'utilisation de l'eau de boisson et étude de la perméabilité osmotique branchiale. J. Physiol. Paris 70 : 605-626.

Kirsch R., Guinier D. (1978). Action of epinephrine and norepinephrine on water, chloride and sodium exchange in the european eel. Gen. Comp. Endocr. 34(1) abstracts 77.

Kirsch R., Meister M.F. (1982). Progressive processing of the ingested water in the gut of sea-water teleosts. J. Exp. Biol. 98, 67-81.

Kirschner L.B. (1978). External charged layer and Na^+ regulation. In "Osmotic and volume regulation". Alfred Benzon Symposium XI. Munksgaard. Academic Press. New York p.310-321.

Kirschner L.B. (1979). Control mechanisms in crustaceans and fishes. In "Mechanisms of osmoregulation in animals". (Gilles R. ed) John Wiley and Sons. p.157-222.

Krogh A. (1939). Osmotic regulation in fresh water fishes by active absorption of chloride ions. Z. vergl. Physiol. 24 : 656-666.

Lahlou B. (1970). La fonction rénale des téléostéens et son rôle dans l'osmorégulation. Bulletin d'Informations Scientifiques et Techniques du Commissariat à l'Energie Atomique. 144 : 17-52.

Lahlou B. (1983). Intestinal transport and osmoregulation in fishes. In "Intestinal transport" (Gilles-Baillien M. and Gilles R.Eds). Springer Verlag. Berlin p.341-353.

Laurent P., Kirsch R. (1975). Modifications structurales de l'oesophage liées à l'osmorégulation chez l'anguille. C.R. Acad. Sci. Paris 280 : 2227-2229.

Leray C., Florentz A. (1983). Biochemical adaptation of trout intestine related to its ion transport properties. Influence of dietary salt and fatty acids, and environmental salinity. In "Intestinal transport" (Gilles-Baillien M. and Gilles R. eds). Springer Verlag. Berlin. p.354-368.

Loretz C.A. (1983). Ion transport by the intestine of the goby, *Gillichthys mirabilis*. Comp. Biochem. Physiol. 75A (2) : 205-210.

Lotan R., Skadhauge E. (1972). Intestinal salt and water transport in a euryhaline teleost, *Aphanius dispar* (Cyprinodontidae). Comp. Biochem. Physiol. 42A : 303-310.

MacKay W.C., Janicki R. (1979). Changes in the eel intestine during seawater adaptation. Comp. Biochem. Physiol. 62A : 757-761.

MacLeod M.G. (1978). Effects of salinity and starvation on the alimentary canal anatomy of the rainbow trout *Salmo gairdneri* Richardson. J. Fish. Biol. 12 : 71-79.

Maetz J. (1971). Fish gills : mechanisms of salt transfer in freshwater and seawater. Phil. Trans. Roy. Soc. Lond. B.262 : 209-249.

Maetz J. (1974). Aspects of adaptation to hypo-osmotic and hyper-osmotic environments. In "Biochemical and biophysical perspectives in marine biology" (Malins D.C. and Sargent J.R. eds.) 1 : 1-167. Academic Press. London.

Maetz J. (1976). Transport of ions and water across the epithelium of fish gills. In "Lung liquids". Ciba Foundation Symposium 38 (New series) Elsevier. Excerpta Medica. North-Holland, Amsterdam. p.133-159.

Mainoya J.R., Bern H.A. (1982). Effects of Teleosts urotensins on intestinal absorption of water and NaCl in Tilapia, *Sarotherodon mossambicus*, adapted to fresh water or sea water. Gen. Comp. Endocr. 47 : 54-58.

Marshall W.S. (1978). On the involment of mucous secretion in teleost osmoregulation. Can. J. Zool. 56 : 1088-1091.

Meister M.F. (1982). Absorption de l'eau ingérée chez les poissons téléostéens : analyse fonctionnelle et structurale. Thèse de doctorat 3è cycle. Université Louis Pasteur. Strasbourg.

Meister M.F., Humbert W., Kirsch R., Vivien-Roels B.(1983). Structure and ultrastructure of the oesophagus in sea water and fresh water teleosts (Pisces). Zoomorphology 102 : 33-51.

Motais R., Isai J. (1972). Temperature dependence of permeability to water and to sodium of the gill epithelium of the eel *Anguilla anguilla*. J. Exp. Biol. 56 : 587-600.

Oide M., Utida S. (1967). Changes in water and ion transport in isolated intestines of the eel during salt adaptation and migration. Marine Biol. 1 : 102-106.

Oide M. , Utida S. (1968). Changes in intestinal absorption and renal excretion of water during adaptation to sea water in the japanese eel. Marine Biol. 1 : 172-177.

Parmelee J.T., Renfro J.L. (1981). Sodium transport across the oesophageal epithelium of a euryhaline marine teleost. Federation Proceedings 40 (3).

Portier P., Duval M. (1922). Pression osmotique du sang de l'anguille (essuyée), en fonction des modifications de salinité du milieu extérieur. C.R. Acad. Sci. Paris 175 : 1105-1106.

Sakata T., Engelhardt W.v. (1981a). Influence of short-chain fatty acids and osmolality on mucin release in the rat colon. Cell.Tiss.Res. 219 : 371-377.

Sakata T., Engelhardt W.v. (1981b). Luminal mucin in the large intestine of mice, rats and guinea pigs. Cell Tiss.Res. 219: 629-635.

Sharratt B.M., Bellamy D., Chester Jones I. (1964). Adaptation of the silver eel (*Anguilla anguilla* L.) to sea water and to artificial media together with observations on the role of the gut. Comp. Biochem. Physiol. 11 : 19-30.

Shehadeh Z.H., Gordon M.S. (1969). The role of the intestine in salinity adaptation of the rainbow trout, *Salmo gairdneri*. Comp. Biochem. Physiol. 30 : 397-418.

Shepard K.L. (1981). The influence of mucus on the diffusion of water across fish epidermis. Physiol. Zool. 54(2) : 224-229.

Shepard K.L. (1982). The influence of mucus on the diffusion of ions across the oesophagus of fish. Physiol. Zool. 55 : 23-34.

Skadhauge E. (1969). The mechanism of salt and water absorption in the intestine of the eel (*Anguilla anguilla*) adapted to waters of various salinities. J. Physiol. 204 : 135-158.

Skadhauge E. (1974). Coupling of transmural flows of NaCl and water in the intestine of the eel (*Anguilla anguilla*). J. Exp. Biol. 60 : 535-546.

Skadhauge E. (1976). Regulation of drinking and intestinal water absorption in euryhaline teleosts. In "Intestinal ion transport" (Robinson J.W.L. ed.) MTP Press. Lancaster p.328-333.

Sleet R.B., Weber L.J. (1982). The rate and manner of seawater ingestion by a marine teleost and corresponding seawater modification by the gut. Comp. Biochem. Physiol. 72A : 469-475.

Smith H.W. (1930). The absorption and excretion of water and salts by marine teleosts. Am. J. Physiol. 93 : 480-505.

Smith H.W. (1932). Water regulation and its evolution in fishes. Quart. Rev. Biol. 7 : 1-26.

Utida S., Oide M., Saishu S., Kamiya M. (1967). Préétablissement du mécanisme d'adaptation à l'eau de mer dans l'intestin ét les branchies isolés de l'anguille argentée au cours de sa migration catadrome. C.R. Sòc. Biol. Paris 161 : 1201.

Utida S., Hirano T., Kamiya M. (1969). Seasonal variations in the adjustive responses to sea water in the intestine and gills of the japanese cultured eel, *Anguilla japonica*. Proc. Jap. Acad. 45 : 293.

Yamamoto M. , Hirano T. (1978). Morphological changes in the oesophageal epithelium of the eel, *Anguilla japonica*, during adaptation to sea water. Cell. Tiss. Res. 192 : 25-38.

Transport properties of the fish urinary bladders in relation to osmoregulation

B.LAHLOU and B.FOSSAT

In teleost fishes, osmoregulation in freshwater (FW) and in sea water (SW) involves the active participation of several organs specialized in water and ion transport. In this context, a component of the excretory system, the urinary bladder, proved to be a convenient and valuable material to study, after Murdaugh et al. (1963) and Lahlou (1967) demonstrated that it modifies urine stored in it, and raised the question of its probable osmoregulatory role.

The present article is a general review dealing with the subject.

The bladder is formed by a distal expansion of the ureters. Therefore, it is embryologically different from that of tetrapods. A few observations have been made concerning its structure. In the trout (our observations) it is made of a single-layered epithelium lining a much thicker muscular wall. The most complete work was done on Gillichtys mirabilis by Nagahama et al. (1975). They found tall (20-50 µm) columnar, mitochondria rich cells, with few microvilli and cilia, low (3-15 µm) cuboidal cells and some basal cells, but no goblet cells. The two most numerous populations, columnar and cuboidal cells, are topographically separated. When fish were transferred from SW to dilute medium (5 % SW), striking changes occurred. In columnar cells, mitochondria developed and formed lamellae while microvilli decreased in number. In other words, these cells appeared to be more active.

I - The bladder as an osmoregulatory organ.

A - In vivo studies.

The easiest way to collect urine in fishes is to insert a cannula in the the bladder through the urinary papilla. Under these conditions, the first fluid obtained is the "bladder urine", initially contained in this organ. This is followed by samples of "ureteral urine" not allowed to be stored in the bladder.

In the early experiments using this protocol, it was found that urine collected from freshly-caught marine fishes was isotonic to blood, rich in magnesium and sulphate, but very low in chloride. Animals kept in the laboratory voided urine in which chloride increased and tended to be equal to its concentration in blood. These changes were considered as reflecting an abnormal state described as "laboratory diuresis".

However from studies on European flounder bearing a permanent cannula in the bladder, Lahlou (1967) showed that these changes in composition depended only on how urine was collected. Each time urine flowing from the ureters was allowed to stay in the bladder for several hours, its Na^+ and Cl^- content decreased. Thus he proposed that the urinary bladder is an additional device of the teleostean osmoregulatory system which completes the osmotic work of the nephrons distally, both in FW and in SW. Similar observations were made in Opsanus tau (Lahlou et al. 1969).

Some time earlier, Murdaugh et al. (1963) reported that inulin and bicarbonate ions introduced into the bladder or injected into the blood of Lophius could cross the wall of the bladder. They suggested that in fishes (and presumably in humans as well) the bladder may modify urine before voiding.

Other in vivo demonstrations of bladder function were made as follows :

- In Lophius americanus and Hemitripterus americanus, Forster and Danforth (1972, 1973) showed by comparing "ureteral" and "bladder" urines that Na^+ and, to a lesser extent, Cl^- were extracted towards the blood. In individuals, Na^+ dropped from 95 to 16 mM and even from 140-180 to 18-48 mM.

- In Salmo gairdnerii, Beyenbach and Kirschner (1975) infused saline or $MgCl_2$ solution into the vascular system in order to provoke diuresis and collected urine from the bladder. They found that Mg^{++} and Na^+ concentrations and excretion were positively correlated. This ruled out the hypothesis presented by Natochin and Gusev (1970) according to which a Mg^{++}/Na^+ exchange takes place across the nephron tubule of Oncorhynchus kisutch and O. nerka. Instead Beyenbach and Kirschner observed that an inverse relationship between Mg^{++} and Na^+ developed inside the bladder because Na^+ was reabsorbed with water, thus increasing Mg^{++} concentration.

These observations also led to the conclusion that the bladder is scarcely permeable to Mg^{++} since the final gradient maintained across it in Lophius was 111 : 0.6 (Forster and Danforth, 1973).

- Howe and Gutknecht (1978) carried out an extensive analysis of bladder and ureteral urines in Opsanus tau. They found that the bladder reabsorbs 60 % of the kidney urine, essentially as a NaCl solution isotonic to plasma. They calculated that the amount of water thus conserved was equivalent to 10 % of the fluid absorbed by the gut.

B - In vitro studies.

In vitro methods were started simultaneously by Hirano et al. (1971) on the starry flounder and by Lahlou and Fossat (1971) on the trout by using the classical sac technique devised by Bentley (1958) for amphibian bladder. This was further applied to the bladders of many other species (Johnson et al., 1972 ; Renfro, 1972 ; Hirano et al., 1973; Hirano, 1975 ; Doneen, 1976 ; Owens et al., 1977).

The perfused sac used by Fossat et al. (1974), Renfro (1975) and Renfro et al. (1976) is essentially similar to the previous technique except that the mucosal fluid is recirculated by an air-lift from an outer reservoir. It is more reliable for ion flux measurements but volume changes produced by the transport of water are too small to be detected.

Ussing chambers were devised for several teleosts, with an exposed area adapted to the size of the bladder : Hemitripterus (Hogben et al., 1972; in split chamber of 1 cm^2), Platichthys stellatus (Demarest, 1977; Demarest and Machen, 1978, 1979 ; area not stated), Salmo irideus (Fossat and Lahlou, 1979 a, b ; 1982 ; 0.6 cm^2), Pseudopleuronectes americanus (Dawson and Andrew, 1979, 1980, 1981 ; 1.25 cm^2), Gillichthys (Loretz, 1979 ; Loretz and Bern, 1980 ; 0.07 or 0.2 cm^2). This method was used to measure short-circuit current or as a voltage-clamp device when no spontaneous electrical potential difference existed across the bladder (e.g. in trout, Fossat and Lahlou, 1982).

The Schultz-type apparatus originally designed for mammalian intestine (Schultz et al., 1967) was applied by Fossat and Lahlou (1979 a) to the trout bladder in order to determine ion fluxes across the apical border of the epithelium. Because of the small size of the bladder, only three separate ports of 0.125 cm^2 each could be made.

Finally, electrochemical gradients across cellular membrane were determined in trout bladder, using conventional and ion-selective microelectrodes (Harvey, 1982 ; Harvey et al., unpublished).

C - Hypertonic transport.

One of the significant results obtained with the sac technique was to show that when the membrane is bathed with identical Ringer on both sides it transfers Na$^+$ and Cl$^-$ in equal amounts from mucosa to serosa. In trout bladder (Lahlou and Fossat, 1971) osmolarity drops in the sac in such a way that an equilibrium is eventually attained in which the sac fluid is diluted up to 10 times compared to the initial Ringer. Moreover decrease in osmolarity is accounted for by the disappearance of Na$^+$ and Cl$^-$ from the sac, indicating that these ions are not exchanged across the bladder wall with others of the same signs. As water moves in the same direction as electrolytes, it is obvious that the membrane transfers a fluid which is highly concentrated in NaCl compared to Ringer. Similar conclusions were reached by Hirano et al. (1971) on Platichthys stellatus. In FW-adapted animals, Na concentration in the passing fluid was 1370 mM as against 150 in Ringer. This hypertonicity is of physiological value in FW fishes since it results in excretion of "free water" in this medium.

This evidence for an osmoregulatory role is supported by other observations in salt media. Hirano et al. found that SW acclimation is followed by a decrease in Na transport (180 mM in passing fluid, which is close to Ringer concentration). Hirano et al. (1973) made an extensive survey of selected fresh-water, marine and euryhaline teleosts. They found that when the bladder is bathed with Ringer it carries less than 5 μl/h.cm^2 in FW-adapted fish ; the value is even nil in Ictalurus. This indicates that FW bladders are relatively impermeable to water. Conversely, a water flow of 12-17 μl/h.cm^2 is observed in SW fish. Increase in water transport in salt media is frequently paralleled by a lower rate of sodium reabsorption. A somewhat different picture was obtained with Gillichthys which may be adapted to salinities ranging from 5 % to 300 % SW and in which Na absorption by the bladder increases in salinities higher than SW (Owens et al., 1977). A common essential feature is that hypertonicity of the transported fluid declines in salt media towards isotonicity. Thus in SW fishes, the bladder removes as much water as possible from its luminal fluid, again a condition favouring osmoregulation.

This adaptation distinguishes the urinary bladder from most other organs which absorb or secrete water and electrolytes by an isotonic transport mechanism (e.g. gallbladder or intestinal mucosa). As in the latter epithelia however, water transport is "solute-linked", that is, it is ensured as long as ions are actively pumped by the membrane. Thus, water flow across trout bladder totally disappears when mucosal Ringer is replaced with isosmotic mannitol (Fossat and Lahlou, 1977). Because of the presence of this mechanism, mucosa-to-serosa osmotic permeability of the bladder can only be estimated when sodium is excluded from the luminal fluid. By this method Fossat and Lahlou (1977) and Fossat (unpublished) found in FW trout the permeability coefficient, P_o, to be 8.3 x 10^{-6} cm/s and hydraulic conductivity, L_p, at 20°C, 3.45 x 10^{-7} cm/s. atm. These values are small compared to those found in other membranes (House, 1974). Besides, trout bladder is able to carry water even against an adverse osmotic gradient (of up to 160 mOsm/l) provided

sodium chloride is present in the mucosal solution. Finally it is important to note that fluid hypertonicity built up by bladders of FW fishes
is one of the highest ever recorded in ion transporting membranes. This
tissue is as efficient as avian salt glands for extracting salt from an
isotonic solution (see House, 1974).

D - Presence of $Na^+ - K^+ - ATPase$

In many cellular and epithelial membranes sodium transport is linked to
an enzyme activity which requires Na^+ and K^+ to split ATP at highest rate. In epithelia, this $Na^+ - K^+$-ATPase is usually localized in the basolateral cell membrane.
     This activity exists in fish bladders. In P. americanus its maxi-
mum in vitro stimulation requires approximately 2 mM K^+ and 30 mM Na^+
in incubation medium (Miller and Renfro, 1973). Ouabain, used as specific inhibitor for $Na^+ - K^+$ -ATPase stops sodium and water transport in
P. stellatus (Johnson et al., 1972), trout (Fossat et al., 1974), P.
americanus (Renfro, 1975). The binding of labelled ouabain demonstrated in electron microscopy that the enzyme was localized in the basal
and lateral membranes (Karnaky et al., 1974) ; Renfro et al., 1976).
     Direct relationship was demonstrated between enzyme activity and
either unidirectional mucosa-to-serosa Na^+ flux (Miller and Renfro,
1973) or, more significantly, net transport of Na^+. In trout, both Na^+
K^+ -ATPase and Na^+ net flux decline simultaneously from FW to one third
SW and to full SW (Fossat et al., 1974 - Figure 1). In P.Americanus in
SW there is also correlation between the two parameters (Renfro et al,
1976). In the euryhaline flounder P. stellatus, Utida et al. (1974) observed a 3-fold increase in enzyme activity from SW to FW.
     These results show that the osmoregulatory capability displayed by
ion and water transport reflects actual adaptation of the cellular enzymatic equipment in fish bladder.

E - Hormonal control.

Some convincing evidence exists to show that the urinary bladder of fishes is under hormonal control as are other osmoregulatory epithelia of
these animals.
     The most original observation was provided by Hirano et al. (1971)
They showed in P. stellatus that prolactin (PRL) injected into SW-adap-
ted fish decreased the in vitro transport of water by the bladder and
brought it down to the level obtained in FW animals, while no effect
was induced in FW-adapted animals. Thus prolactin appears to specifically reduce water permeability. Also, in Platichthy flesus, Hirano (1975)
found that bladders of PRL-treated SW fish reabsorbed NaCl at a greater
rate than did control bladders. PRL is frequently described as the most
versatile hormone in its effects in vertebrates. Its cellular action involves long term changes, cell proliferation and stimulated protein synthesis. Hirano et al. (1973) demonstrated in the flat fish Kareius bico-
loratus that PRL increased ^3H - thymidine incorporation into the DNA
fraction of the urinary bladder. Moreover, Utida et al. (1974) showed
that injection of PRL into SW P. stellatus induces a stimulation of Na^+
- K^+ - ATPase activity of the bladder, simulating the results of FW
transfer.
     More recently, other hormonal substances were assayed. Johnson
(1973) suggested that cortisol and PRL may act as antagonists to each
other. In Gillichthys, Doneen (1976) showed that cortisol stimulates
Na^+ absorption and elevates water permeability in SW fish. Finally, Doneen and Bern (1974) showed that cortisol was required to maintain elevated water permeability in organe-cultured bladder of Gillichthys mirabilis.

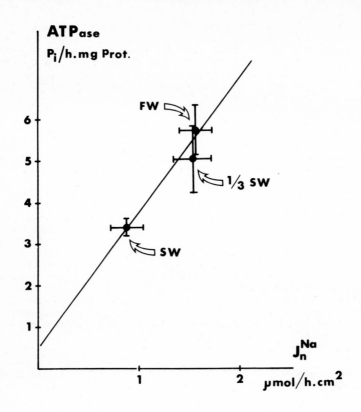

Figure 1
Parallel changes of total ATPase activity and net flux of sodium in trout urinary bladder in relation to external salinity. Animals were adapted to freshwater (FW), one-third sea-water (1/3 SW) or full sea-water (SW).
Figure drawn from data of Fossat et al. (1974) ; variations in ATPase are accounted for by only those of Na^+ - K^+ - ATPase ; means $\pm$ standard errors, $n \geq 7$ determinations for each average value of sodium flux or enzyme activity.

Strikingly, fish urinary bladder is not sensitive to neurohypophyseal peptides, arginine-vasotocin in particular (Johnson et al., 1972)
This is at variance with the rapid and large effects thoroughly described in anuran bladder. This lack of action may be a consequence of the
mesonephric origin of fish bladder or may indicate that cellular recep
tors to neurohypophyseal hormones have evolved with adaptation to terrestrial life in vertebrates.

II - The bladder as a permeable membrane : leaky or tight ?

Epithelial membranes are currently described as tight or leaky depending
on their electrical properties and correlated transport mechanisms.

A - Electrical measurements .

For fish urinary bladders the available data on electrical parameters
measured in Ussing chambers are presented in Table 1. As a rule, the
transepithelial potential differences (PDs) are in the lower range of
values recorded in other tissues and may be serosa positive or negative.
The bladder may even be electrically silent, as in trout bladder (Lahlou and Fossat, 1971 ; Fossat and Lahlou, 1979 a) in all circumstances
in which it is not faced with a transmural chemical gradient.
    The electrical resistance of the tissue is strikingly variable. In
trout bladder, values are in the range of 100-300 $\Omega.cm^2$. As the PD is
zero, this tissue compares relatively well with the gallbladder and we
have therefore described it as "moderately leaky". Much higher values
(up to 1800 $\Omega.cm^2$) were obtained however in other species, all living
in SW. These compare with the resistance of tight epithelia, such as
frog skin and bladder. Another peculiarity is that resistance may chan
ge with the outer salinity to which the fish are adapted. Thus, it increases in salt water trout (our observations). In general,it is higher
in SW than in FW animals. However, in Gillichthys mirabilis, transfer
to dilute media induces an increase in resistance by about 5-fold in the
columnar cells region (Loretz and Bern, 1980).
    This large scatter is surprising in as much as the transport pro-
perties of fish bladders are relatively stable from species to species.
Two possible artifacts at least may explain it. One is that the degree
of stretching is likely to result in variable electrical resistance and
edge-damage. Another error may arise because the bladder is rarely free
in the body cavity and is more or less uneasy to separate from adjacent
layers. As these tissues are probably inert as far as ion transport is
concerned, they are likely to provide an additional electrical resistance without changing the actual PD recorded across the bladder.
    From available data, it is therefore difficult to classify all fish
bladders as leaky or tight epithelia. In our hands,electrical measurement made on trout (in FW)and European flounder and plaice (in SW) yiel-
ded relatively low values for the resistance (less than 500 $\Omega.cm^2$) and
either no potential or else small potential (less than 5 mV). By contrast the bladders of G. mirabilis (Loretz and Bern, 1980) and of P.ame-
ricanus (Dawson and Andrew, 1979) are held as tight epithelia. If reported measurements are all valid, then it appears that change in fish a-
adaptation from FW to SW will result in a large shift from leaky to
tight epithelium in the bladder, a fact not described in vertebrate epithelia.

B - Neutral or electrical coupling in NaCl transport ?

In gallbladder, Diamond (1962) concluded that active absorption of Na^+
and Cl^- was the result of a coupled, electrically neutral process.

TABLE 1. Transepithelial electrical measurements reported in fish urinary bladders.

Species	Adaptation Medium	PD (serosa) mV	Resistance $\Omega.cm^2$	Reference
Hemitripterus americanus	SW	-2.26	500 - 1000	Hogben et al., 1972
Platichthys stellatus	SW	0.2	500	Demarest, 1977
" "	FW	8.5	340	"
Pseudopleuronectes americanus	SW	-1.3	850	Renfro, 1975
"	SW	-	1600	Dawson and Andrew, 1979
Gillichthys mirabilis				
columnar cells	**SW**	14	287	Loretz and Bern, 1980
	SW 5%	4.6	1780	"
cuboidal cells	SW	2.8	1280	"
	SW 5%	2.4	1720	"
Opsanus tau (in vivo)	SW	5	-	Howe and Gutknecht, 1978
Salmo gairdnerii	**FW**	0	100 - 300	our observations
"	SW	-	415	"
Platichthys flesus	SW	-1 to -5	430	"
"	FW	-1 to -8	-	"

The existence of this mechanism has since been proposed for the apical or basal membrane of other epithelia.

In fish urinary bladder, the occurrence of such a coupling is supported by the finding that Na^+ or Cl^- transport is inhibited partially in P. americanus (Renfro, 1977), completely in trout (Fossat and Lahlou, 1977, 1979a) when the co-ion is replaced by a non-transported ion (such as choline and gluconate respectively) in the mucosal solution or in both sides.

Bearing in mind that all NaCl transporting epithelia possess a ouabain sensitive Na^+ - K^+ - ATPase for which there is no compelling evidence for Cl^- requirement, one may propose that both identified processes , i.e. neutral coupled transport and electrical coupling, are likely to exist together in these tissues.

By using the Schultz-type apparatus for trout bladder, we could show that Na^+/Cl^- coupling takes place at the apical border of the epithelium (Fossat and Lahlou, 1979a). Relationships between Na^+ and Cl^- entry and their respective mucosal concentrations obey saturation kinetics, indicating that some carrier-mediated mechanism of high affinity (Km = 8 mM) operates in the membrane. Moreover the apical uptake of these ions appears to be the limiting step for transepithelial transport since its amplitude is equal to the mucosa-to-serosa fluxes, except for an exchange-diffusion component which accounts for part of the chloride exchanges.

A slightly different picture prevails in bladders which display a consistent transepithelial PD. The above coupled process accounts for only one part of the ionic uptake. In P. americanus about 50 % of Na^+ or Cl^- transport remains when its co-ion is completely removed from the mucosal solution (Renfro, 1977). In P.stellatus, coupled transport represents 40 % of mucosa-to-serosa ion fluxes (Demarest , 1977; Demarest and Machen, 1978, 1979). Independent transport of Na^+ and Cl^-, as kwown to occur in crustacean and fish gills and in amphibian membranes in particular, may serve other physiological needs than osmoregulation. For example Na^+/H^+ or NH_4^+ and Cl^-/HCO_3^- exchanges may contribute to acid-base regulation and to CO_2-nitrogen excretion. Renfro (1975) indicated that flounder bladder acidifies urine and that its H^+ secretion is greater than that seen in toad bladder. In our experiments on trout (unpublished) addition of acetazolamide to the serosal fluid or removal of HCO_3^- ions from the bathing solutions did not alter chloride transport. These peculiarities may also be disclosed by using drugs which interfere with specific mechanisms. Amiloride which acts on Cl^--independent Na^+ sites reduces $J_{ms}Na^+$ in P.americanus, while it is inefficient in trout (Renfro, 1977). Conversely it was observed in American flounder that furosemide and ethacrynic acid which act on the Na/Cl coupled moiety inhibit ion water transport.

In conclusion, fish bladders present an electrically silent NaCl transport on which an electrogenic ionic transfer may be surimposed, as revealed by the recording of short-circuit current. Where the latter exists however, its ionic basis remains to be clearly defined as it is not equated with Na^+ transport (see below).

C - Paracellular permeability.

In leaky epithelia, the low electrical resistance and PD are ascribed to the presence of shunt pathways of high ionic conductance located in the paracellular spaces. Because of the presence of fixed negative charges, the apical junctions are cation-selective. Consequently much of the Na^+ transported into the lateral spaces may recycle towards the mucosal solution. Backflux of Na^+ and forward flux of Cl^- will create a diffusion potential which opposes and may reverse the serosa-positive PD produced by the activity of the basolateral Na^+/K^+ pump. This mechanism has been emphasized by Field et al. (1978) to explain the serosal the serosal electronegativity depicted in intestinal mucosa of marine fishes.

This model may well apply to fish urinary bladder, as it may account suitably for the various conditions encountered for PDs (Table 1). When studying trout bladder, we were struck by the finding that the transepithelial fluxes of chloride far exceeded those of sodium, contrary to what is most frequently observed in epithelia. Therefore we addressed the question of whether the tight junctions of fish bladders are cation-selective or not. For this purpose, ion fluxes were measured while tissue was submitted to voltage-clamp from 0 to $\pm$ 50 mV, and were plotted according to the method of Schultz and Zalusky (1964) which permits discrimination between the transcellular and paracellular pathways. For each potential imposed, experiments were performed under control conditions or in the presence of 2, 4, 6 - Triaminopyrimidine (TAP) which elicits a reversible increase in tissue resistance due to a specific decrease in Na^+ permeability of the paracellular pathway (Fossat and Lahlou, 1979b and Figure 2). This study showed firstly that the diffusional, potential-dependent flux of Cl^- is higher than that of Na^+, secondly that TAP has no effect on this tissue. We therefore concluded that the paracellular pathway is not cation-selective in trout bladder in current conditions.

In P. stellatus, Demarest and Machen (1978) measured total conductance (G_t) of the bladder in relation to salinity adaptation. They found that FW bladders exhibit a G_t ten times greater and fluxes two to five times greater than SW bladders. From analysis of voltage-flux relationships they concluded that osmoregulatory adjustments in the bladder involves, at least in part, changes in the apical cell membranes, with increased sodium conductance.

D - Ionic channels in cell apical membrane.

Electrochemical potentials across the apical membrane have been measured for the first time in a fish urinary bladder by Harvey (1982) and Harvey et al. (unpublished). Conventional and ion-sensitive microelectrodes were applied to the trout bladder. In the presence of Ringer, the transepithelial potential was zero (as expected in this species)and the intracellular potential was -56 mV. The intracellular ionic activities of sodium, potassium and chloride (A_{Na}^i : 16 mM ; A_K^i : 76 mM ; A_{Cl}^i: 21 mM) indicated an active accumulation of K^+ and Cl^- and an active extrusion of Na^+ by the cell. The paracellular conductance accounted for about 90 % of transepithelial current flow and thus confirmed leakiness of the tissue.

When trout bladder is submitted to the effect of the polyene antibiotic Amphotericin B on it apical side, there is an immediate serosapositive potential, the amplitude of which is related to the total conductance of the tissue (Fossat and Lahlou, 1982 ; Harvey, 1982 ; Harvey et al., unpublished). From their analysis, Harvey et al., showed that its main effect was to open Na^+ and K^+ channels in the apical cell membrane, while Cl^- permeability was not increased. In addition, the induced potential was abolished in the presence of Ba^{++} ions in the mucosal medium, an effect revealing occurence of K^+ channels at that side.

Polyene antibiotics have also been used by other authors. In P. stellatus, Demarest and Machen (1979) briefly reported that nystatin increased total conductance in SW but not in FW bladders. In SW P. americanus, Dawson and Andrew (1980, 1981) showed in a series of experiments in which they used Amphotericin B, ouabain and varying concentrations of K^+ in bathing media, that the bladder contains in its apical membrane a potassium channel through which potassium is secreted towards the mucosal side. Moreover, the short-circuit current recorded under control conditions in this tissue (which, unlike trout bladder, presents a spon-

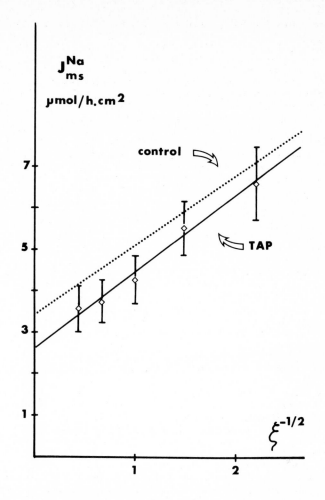

Figure 2
Mucosa-to-serosa unidirectional flux of sodium (J_{ms} Na) plotted as a
function of externally applied potential, according to the method of
Schultz and Zalusky (1964), in trout urinary bladder mounted in Ussing
chambers (see text).
Dotted line corresponds to control bladders as results from a large number of experiments. Solid line represents measurements in the presence
of TAP 20 mM added to the mucosal medium, pH 6.1 ; means ± standard errors, n=6. The absence of significant difference in slope and in verti-
cal intercept of the two regression lines indicates that TAP has no effect on potential-dependent (presumably paracellular) and potential-
independent (non diffusional, transcellular) sodium fluxes.

taneous PD) is accounted for by its potassium transport. This is an unusual and interesting result which provides the awaited the awaited ionic basis of membrane polarization (referred to above). In addition, it is in line with recent developments showing that in other epithelia, such as fish intestine, ion transport involves in various ways the triple association of Na^+, Cl^- and K^+ ions at the apical and basolateral membranes (Musch et al., 1982).

Although potassium secretion is quantitatively small compared to sodium and chloride transport, its implications in the present case may go beyond the simple electrochemical mechanisms of cell membranes. Namely it may intervene in osmoregulation as well. In the aglomerular fish, Opsanus tau, it was occasionally observed (Lahlou, unpublished) that "bladder" urine, unlike "ureteral" urine collected simultaneously, contained high K^+ (100 mM) and low Na^+ concentrations, a condition not explained by water reabsorption alone.

The present work was aided by the Centre National de la Recherche Scientifique (E.R.A. 943)

References

Bentley PJ (1958) The effect of neurohypophyseal extracts on water transfer across the wall of the isolated urinary bladder of the toad Bufo marinus. J Endocr 17:201-209.

Beyenbach KW and Kirschner LB (1975) Kidney and urinary bladder functions of the rainbow trout in Mg and Na excretion. Am J Physiol 229: 389-393.

Dawson DC and Andrew D (1979) Differential inhibition of NaCl absorption and short-circuit current in the urinary bladder of the winter flounder, Pseudopleuronectes americanus. Bull Mt Desert Isl Biol Lab 19:46-49.

Dawson DC and Andrew D (1980) Active potassium secretion by flounder urinary bladder : role of a basolateral Na-K pump and apical potassium channel. Bull Mt Desert Isl Biol Lab. 20:89-92.

Dawson DC and Andrew D (1981) Basolateral potassium conductance in flounder urinary bladder. Bull Mt Desert Isl Biol Lab 21:29-31.

Demarest JR (1977) Freshwater acclimation in the teleost urinary bladder : changes in transport (Abstract) Am. Zool 17:877

Demarest JR and Machen TE (1978) Minor contribution of paracellular transport in the euryhaline teleost urinary bladder (UB) (Abstract) Fed Proc 37:511.

Demarest JR and Machen TE (1979) Active Na and Cl transport by the euryhaline teleost urinary bladder (UB) (Abstract) Fed Proc 38:1059.

Diamond JM (1962) The mechanism of solute transport by the gallbladder. J Physiol London 161:474-502.

Doneen BA (1976) Water and ion movements in the urinary bladder of the Gobiid teleost Gillichthys mirabilis in response to prolactin and to cortisol. Gen Comp Endocrin 28:33-41.

Doneen BA and Bern HA (1974) In vitro effects of prolactin and cortisol on water permeability of the urinary bladder of the teleost Gillichthys mirabilis. J exp Biol 187:173-179.

Field M , Karnaky K J Jr, Smith, PL , Bolton, JE and Kinter, W.B. (1978) Ion transport across the isolated intestinal mucosa of the winter flounder Pseudopleuronectes americanus. I - Functional and structural properties of cellular and paracellular pathways for Na and Cl. J Membrane Biol 41:265-293.

Forster R P and Danforth JW (1972) Osmoregulatory role of the urinary bladder in the stenohaline marine teleost Lophius americanus and Hemitripterus americanus. Bull Mt Desert Isl Biol Lab 12:35-37.

Forster RP and Danforth JW (1973) Transport of fluid and electrolytes by urinary bladder of the aglomerular marine teleost Lophius americanus Bull Mt Desert ISL Biol Lab 13:42-44.

Fossat B and Lahlou B (1977) Osmotic and solute permeabilities of isolated urinary bladder of the trout. Am J Physiol 233:F525-531.

Fossat B and Lahlou B (1979a) The mechanism of coupled transport of sodium and chloride in isolated urinary bladder of the trout. J Physiol (London) 294:211-222.

Fossat B and Lahlou B (1979b) Failure of 2,4,6 -Triaminopyrimidine to block sodium pathways in a "leaky" epithelium : the urinary bladder of the trout. Pfluegers Archiv, 379:287-290.

Fossat B and Lahlou B (1982) Ion flux changes induced by voltage-clamping or by Amphotericin B in isolated urinary bladder of the trout. J Physiol London 325:111-123.

Fossat B, Lahlou B and Bornancin M (1974) Involvement of a Na-K ATPase in sodium transport by fish urinary bladder. Experientia 30:376-377.

Harvey BJ (1982) Microelectrode determination of electrochemical potentials in epithelia. PhD Thesis Dublin.

Harvey BJ, Kernan RP and Lahlou B (in preparation) Electrochemical potentials in trout urinary bladder : effects of Amphotericin B.

Hirano T (1975) Effects of prolactin on osmotic and diffusion permeability of the urinary bladder of the flounder Platichthys flesus. Gen Comp Endocrin 27:88-94.

Hirano T, Johnson DW and Bern HA (1971) Control of water movement in flounder urinary bladder by prolactin.Nature 230:469-471.

Hirano T, Johnson DW, Bern HA and Utida S (1973) Studies on water, ions movements in the isolated urinary bladder of selected freshwater, marine and euryhaline teleosts. Comp Biochem Physiol 45:529-540.

Hogben CAM, Brandes M, Danforth J and Forster RP (1972) Ion transport by the isolated urinary bladder of a teleost Hemitripterus americanus. Bull Mt Desert Isl Biol Lab 12:52-57.

House CR (1974) Water transport in cells and tissues. Arnold, E.Ltd Ed London.

Howe D and Gutknecht J (1978) Role of urinary bladder in osmoregulation in marine teleost, Opsanus tau. Am J Physiol 235:R48-54.

Johnson DW (1973) Endocrine control of hydromineral balance in teleosts. Am Zool 13:799-818.

Johnson DW, Hirano I, Bern HA and Conte FP (1972). Hormonal control of water and sodium movements in the urinary bladder of the starry founder, Platichthys stellatus. Gen Comp Endocrin 19:115-128.

Karnaky KJ, Renfro JL, Miller DS, Church HH and Kinter WB (1974) Teleost urinary bladder (winter flounder, Pseudopleuronectes americanus) : ultrastructure and Na/K-ATPase localization. Bull Mt Desert Isl Biol Lab 14:47-51.

Lahlou B. (1967) Excrétion rénale chez un poisson euryhalin, le flet (Platichthys flesus L.). Caractéristiques de l'urine normale en eau douce et en eau de mer et effets des changements de milieu. Comp Biochem Physiol 20:925-938.

Lahlou B and Fossat B (1971) Mécanisme du transport de l'eau et du sel à travers la vessie urinaire d'un poisson téléostéen en eau douce, la truite arc-en-ciel. CR Acad Sci 273, 2108-2110.

Lahlou B., Henderson IW and Sawyer WH (1969) Renal adaptation by Opsanus tau a euryhaline aglomerular teleost to dilute media. Am J Physiol 216:1266-1272.

Loretz CA (1979) Electrolyte transport by the urinary bladder of a euryhaline teleost (Abstract) Am Zool 19:944.

Loretz CA and Bern, HA (1980) Ion transport by the urinary bladder of the Gobiid teleost, Gillichthys mirabilis. Am J Physiol 239:R415-423.

Miller DS and Renfro JL (1973) Preliminary studies on the characterization of Na-K-ATPase and its relationship to active Na transport in the urinary bladder of the winter flounder, Pseudopleuronectes americanus. Bull Mt Desert Isl Biol Lab 13:80-82.

Murdaugh HV, Soteres P, Pyron W and Weiss E (1963) Movement of inulin and bicarbonate ion across the bladder of an aglomerular teleost, Lophius americanus. J Clin Invest 42:959.

Musch MW, Orellana SA, Kimberg LS, Field M, Halm DR, Krasny EJ Jr and Frizzell RA (1982) $Na^+-K^+-Cl^-$ co-transport in the intestine of a marine teleost. Nature 300:351-353.

Nagahama Y, Bern HA, Doneen BA and Nishioka RS (1975) Cellular differenciation in the urinary bladder of a euryhaline marine fish Gillichthys mirabilis, in response to environmental salinity change. Development, Growth and Differentiation 17:367-381.

Natochin Yu V and Gusev GP (1970) The coupling of magnesium secretion and sodium reabsorption in the kidney of teleosts. Comp Biochem Physiol 38:107-111.

Owens A, Wigham T, Doneen B and Bern HA (1977) Effects of environmental salinity and hormones on urinary bladder function in the euryhaline teleost, Gillichthys mirabilis. Gen Comp Endocr 33:526-530.

Renfro JL (1972) Water and ionic movements across the isolated urinary bladder of the winter flounder Pseudopleuronectes americanus. Bull. Mt Desert Isl Biol Lab 12:81-85.

Renfro JL (1975) Water and ion transport by the urinary bladder of the teleost Pseudopleuronectes americanus.Am J Physiol 228:52-61.

Renfro JL, Miller DS, Karnaky KJ and Kinter WB (1976) Na-K-ATPase localization in teleost urinary bladder by ^3H – ouabain autoradiography. Am J Physiol 231:1735-1743.

Renfro JL (1977) Interdependence of active Na^+ and Cl^- transport by the isolated urinary bladder of the teleost, Pseudopleuronectes americanus. J Exp Zool 199:383-390.

Schultz SG, Curran PF, Chez RA and Fuisz RE (1967) Alanine and sodium fluxes across mucosal border of rabbit ileum. J Gen Physiol 50:1241-1260.

Schultz SG and Zalusky R (1964) Ion transport in isolated rabbit ileum I - Short circuit current and Na fluxes. J Gen Physiol 47:567-584.

Smith HW (1930) The absorption and excretion of water and salts by marine fishes. Am J Physiol 93:480-505.

Utida S, Kamiya M, Johnson DW and Bern HA (1974) Effects of freshwater adaptation and of prolactin on sodium-potassium-activated adenosine triphosphatase activity in the urinary bladder of two flounder species. J Endocrin 62:11-14.

The contrasting roles
of the salt glands, the integument and behavior
in osmoregulation of marine reptiles

W.A. DUNSON

I. INTRODUCTION

Saline waters are a rigourous environment for reptiles in the sense that very few species are able to cope with the problems of hypoosmoregulation. Extracellular fluid concentration is maintained at approximately the same level in reptiles living in deserts, fresh water, or in the sea (Dunson, 1979a; Minnich, 1979, 1982). A general trend among the reptiles is for development of extrarenal or extra-cloacal mechanisms to handle electrolyte excretion under dehydrating conditions on land or in salt water. This is a consequence of the ina-bility of the kidney to secrete urine hyperosmotic to the plasma (Dantzler, 1976). Ever since the discovery in 1958 by Schmidt -Nielsen and Fänge of salt glands in reptiles, attention has focused on the role of these unique organs in osmoregulation (Dunson, 1976, 1981a; Peaker and Linzell, 1975). The presence of salt glands in all fully marine reptiles and their striking functional similarity to avian nasal salt glands has perhaps contributed to a misunderstanding of their actual role in water and ion balance. While salt glands seem quite important as osmoregulatory organs, they are only part of an overall system that regulates both intake and loss of salts and water. The first indication that organs other than the salt glands were critical factors in osmoregulation of marine reptiles came from studies on variation in excretion rate among sea snake salt glands (Dunson and Dunson, 1974). There was a remarkable amount of variation among species with similar feeding habits. This led to a search for concomitant variation in intake rates of water and ions. The skin was the most likely candidate since it serves as the primary barrier regulating

the entry of materials from sea water. Subsequent studies of integumentary permeability have shown that some sea snakes do have a considerably greater permeability (to water especially) than others
(Dunson, 1978; Dunson and Robinson, 1976; Dunson and Stokes, 1983).
It is still unclear why there is any advantage for a sea snake to
have a skin more permeable to water. It is now known that even fully
marine reptiles can osmoregulate quite well with only tiny salt glands,
and that skin permeability might be associated in some way with variations in salt gland size and thereby excretion rate. Thus the stage
is set for recognition of the diverse mechanisms by which marine
reptiles osmoregulate. This should have been obvious earlier due to
the many different reptilian lineages that have given rise to marine
and estuarine species, and the consequent scope for evolutionary
variation in adaptational strategies. We are now also able to seek
an answer to the basic question of what specific adaptations a fresh
water reptile actually needs to become marine. Studies of the diversity of modern estuarine forms that are in the process of invading
saline waters offer many significant clues.

The purpose of this paper is to examine the origin of marine
reptiles by carefully studying a series of cases of reptiles that
show a gradation in their ability to hypoosmoregulate in sea water.
This treatise will also be a plea for a more holistic approach to the
study of osmoregulation. No single organ system operates in isolation
and it is necessary for us to attempt to integrate all components
of the overall osmoregulatory scheme to properly understand the striking adaptations of marine reptiles.

II. Types of Salt Glands

Marine reptiles are of three general kinds, turtles, crocodilians,
and squamates (lizards and snakes). These subdivisions represent
different evolutionary lines with no common marine ancestor. Thus all
special adaptations for marine life have arisen independently at least
three times. This fact is well illustrated by the five nonhomologous
salt glands found in the three groups (Table 1). Note however that
the lizards have developed a nasal salt gland, homologous with that
of birds, but independently derived. Within the three major reptilian
groups, there is further evidence for independent derivation of salt
glands. If one considers that each family represents a homogeneous
phylogenetic lineage, then the possession of identical lachrymal

glands by three different families of turtles represents parallel evolution.

TABLE 1 . The adaptive radiation of reptiles into saline waters.

Family	Generic Example	Marine (M), Estuarine (E) or Fresh Water (F)	Salt Gland
Turtles :			
Dermochelyidae	*Dermochelys*	M	Lachrymal
Cheloniidae	*Chelonia*	M	Lachrymal
Emydidae	*Malaclemys*	E	Lachrymal
Kinosternidae	*Kinosternon*	E, F	None
Chelydridae	*Chelydra*	E, F	None
Lizards :			
Iguanidae	*Amblyrhynchus*	M	Nasal
Varanidae	*Varanus*	E	Nasal
Snakes :			
Hydrophiidae	*Pelamis*	M	Sublingual
Acrochordidae	*Acrochordus*	M, E, F	Sublingual
Colubridae	*Cerberus*	M, E, F	Premaxillary
Colubridae	*Nerodia*	E	None (?)
Crocodiles :			
Crocodylidae	*Crocodylus*	E, F	Lingual

Two families of snakes have independently developed posterior sublingual salt glands, but a third possesses a new type , the premaxillary (Table 1). The crocodilians, the last remnants of the archosaurs (ruling reptiles), possess a fifth type of salt gland, the lingual. Yet among all the marine birds, only a nasal salt gland is found. This great variety of reptilian salt gland types should immediately make us cautious in generalizing about their function. Yet it is not uncommon to find all avian and reptilian salt glands lumped as a group in textbooks, despite their diverse origins. There is also a fundamental ultrastructural difference between these two types of glands; reptilian salt glands do not have basal cell infoldings as do those of birds (Dunson, 1976).

III. The Marine Turtles

The Dermochelyidae are represented by a single species of the genus *Dermochelys* (Leatherback turtle). This form is very little known since the adults are extremely large, pelagic, and rare. Apparently they feed primarily on coelenterates and must thereby ingest large quantities of salts. The huge lachrymal gland is assumed to be a salt gland, but has never actually been studied. The shell is covered with a unique soft integument whose permeability characteristics are unknown. Any new knowledge of the osmoregulatory physiology of this unusual creature would be of considerable interest.

The Cheloniidae include about a dozen species of typical sea turtles, such as the green (*Chelonia*) and loggerhead *(Caretta)* turtles. They have a rather normal although reduced shell, and consist of herbivorous and carnivorous species. Very little is understood about overall pathways of water and electrolyte exchange (such as the role of the integument and drinking); the salt gland is large in the few cases studied.

The only marine/estuarine representative of the family Emydidae is the diamondback terrapin, *Malaclemys*. This remarkable turtle is restricted to coastal or estuarine areas and is the only turtle lacking flippers to possess a salt gland. It is most closely related to fresh water turtles of the *Graptemys* group and seems little modified externally for a marine life. However it has extensive osmoregulatory adaptations and is perhaps the most thoroughly studied of any aquatic reptile (see Robinson and Dunson, 1976, and reviews by Dunson, 1979a, and Minnich, 1979, 1982). It is in a very interesting phase evolutionarily since it probably represents an advanced transitional phase to pelagic life passed through by the above two families of true sea turtles.

The final two turtle families from Table 1, the Kinosternidae (mud turtles) and Chelydridae (snapping turtles) consist of predominantly fresh water species. They contain a few carnivorous forms that show an early stage in the evolution of estuarine adaptations. For this reason they are extremely interesting despite the present lack of much physiological data. These turtles live in brackish waters of island ponds and in the upper tidal reaches of coastal creeks.

IV. The Marine Lizards and Snakes

Lizards are not well represented in saline habitats. The famous Galápagos marine iguana (*Amblyrhynchus*) feeds under the sea surface on algae, but spends most of its time basking on coastal rocks. It is a representative of a widespread family (Iguanidae) that contains numerous terrestrial hervivorous and carnivorous forms with nasal salt glands. Thus the evolutionary route by which some of its osmoregulatory adaptations arose is quite different from the terrestrial to fresh water to estuarine pathway apparently followed by all other marine reptiles. There are a series of lizards in other families, exemplified by *Varanus*, which feed in coastal swamps or on beaches and also utilize a nasal salt gland presumably derived from terrestrial ancestors. While many of these lizards can swim, most of their feeding appears to take place on exposed intertidal areas at low tide.

The snakes are by far the most successful and numerous marine reptiles. There are about 50 to 60 species in three to four families (depending on the taxonomic scheme used) that are abundant in coastal

to open sea habitats. Some of the sea snakes are the only living totally pelagic reptiles, that complete their entire life cycles at sea.
The majority of the marine serpents belong to the family Hydrophiidae.
These are essentially sea dwelling cousins of the cobras (Elapidae)
that have radiated extensively in shallow tropical seas of the western
Pacific and Indian Oceans. Most feed on fish and all are venomous. One
branch of this group, the banded sea snakes or sea kraits (*Laticauda*),
is more amphibious and appears to represent a separate lineage from
the elapid stock. Some authorities place them in the Elapidae. Another
old world family, the Acrochordidae, contains only two species, one of
which lives both in fresh water and in the sea (*Acrochordus granulatus*).
Simultaneous occurrence in both habitats is quite unusual, although
the dog-faced water snake (*Cerberus*, Colubridae) shares this trait.
Both *Cerberus* and *Acrochordus* are highly adapted for life in saline
waters, although their salt glands are small. A much more recent
immigrant to coastal waters is found in the new world *Nerodia fasciata*
group of subspecies (*compressicauda, clarki, taeniata*). They may lack
a salt gland, but are quite tolerant of immersion in salt water. Since
they possess a much greater degree of physiological adaptation to sea
water than *Kinosternon* or *Chelydra*, but less than *Malaclemys*, they
represent important transitional forms in the recreation of the presumed sequence of evolution of marine forms from their fresh water pro-
genitors.

V. The Marine Crocodilians

      Crocodylus porosus and *C.acutus* both possess lingual salt glands,
but *C.porosus* seems to be better adapted for osmoregulation in saline
habitats (Grigg et al., 1980). Recently hatched *C.acutus* are quite
susceptible to dehydration and hypernatremia and may require exogenous
fresh water for drinking (Dunson, 1982; Mazzotti and Dunson, submitted).
These two crocodiles are basically coastal mangrove swamp or riverine
animals that only rarely go into the open sea. They do, however, live
inland in fresh water habitats in certain localities.

The True Sea Snakes : The Ultimate in Marine Adaptations

      It is difficult to determine exactly why the hydrophiid sea
snakes are so successful, in contrast with other marine reptiles. The
two other ophidian families containing marine species have only a few
marine members. Osmoregulatory adaptations must be a critical necessity
for any reptile invading marine environments. This is so because all

reptiles maintain approximately the same extracellular fluid concentration (150-200 mM sodium) and the ability to hypoosmoregulate is not
common. This latter inability is usually considered to be mainly a
renal problem, but the death of fresh water snakes in sea water seems
unrelated directly to urine concentrating capacity (Dunson, 1978).
Fresh water snakes placed in sea water usually undergo a high sodium
influx (due to leakage in the oral region), which leads to hypernatremia, massive drinking , and death. Thus one of the most important
adaptations of sea snakes is for a tight sealing of orifices, accompanied by a relatively impervious skin. The lips, nares, and the cloacal opening of sea snakes are indeed closed tightly, and the keratini-
zed integument blocks intake of salts (Dunson, 1975, 1978, 1979a).
Additionally it is likely that there is an internal homeostatic
response to dehydration that inhibits drinking of water above a certain salinity.

The essential impermeability of sea snake integument to sodium,
discovered by Dunson and Robinson (1976), was probably the first
of a series of adaptations developed for life in saline waters. Snakes
have an inherent disadvantage in that their body shape presents a
large surface area relative to their mass (Dunson, 1978). One might
also suppose that a relatively low degree of integumentary water permeability would be advantageous, but this is only partly the case. Most
sea snakes have a fairly low rate of integumentary water exchange ,
but a few species are much higher (Dunson, 1978; Dunson and Stokes,
1983; Stokes and Dunson, 1982a). Higher integumentary water fluxes
are also associated with a slightly greater degree of electrolyte
permeability. The large variability in skin water permeabilities of
highly evolved sea snakes remains an enigma. It is possible that
variations in skin gas transport might be related to this issue, but
adequate date are lacking. The importance of the integument as a route
for water exchange is paramount in all sea snakes. Even in *Pelamis*, a
species with low absolute rates of integumentary water flux (40-60 μ
moles/cm^2.h), virtually all body water turnover occurs through the
skin (Dunson, 1979a).

The role of the integument in osmoregulation may be more sophis-
ticated than acting as a simple barrier. Recent evidence suggests that
downhill diffusion of water and salts across the keratin sheet is
unequal when the chemical gradient remains the same and the skin is
flipped over (Dunson and Stokes, 1983). Of course all net diffusion
across the dead keratin is with the chemical gradient. Yet it appears

that some mechanism in the keratin polarizes or makes the diffusion
asymmetric in a direction that favors hypoosmoregulation by the intact
snake (Table 2). Downhill diffusion of water from the outside (pure
water) to the inside (1 M sodium chloride) of the skin of sea snakes
is faster than when the skin is reversed in the chamber with no change
in the position of the fluids. In a fresh water snake the direction of
polarization of water movement is in the opposite direction (Table 2),
even though the same solutions are used. If similar phenomena operate
in vivo, the net diffusion of water out of marine snakes and into
fresh water snakes could be slowed, contributing to osmoregulatory
homeostatic mechanisms appropriate for each environment. A similar
effect is observed for ion movements (Table 2). Diffusion of sodium
and chloride across a marine snake skin is more rapid in an outward
than in an inward direction; the process is reversed in a fresh water
species (at least for bromide). Here again the net gain of salts in
sea water and net loss in fresh water might be slowed by such processes in vivo, to the benefit of ionic regulation of the body fluids.
These data are still preliminary since they need to be repeated under
conditions that control for all possible artifacts. However the process
of asymmetric diffusion is an exciting new concept in osmoregulation of
aquatic reptiles.

      The skin serves as a passive mechanism of restricting diffusion
of water and solutes between a snake and its environment. Salt glands,
in contrast, actively excrete electrolytes against the prevailing chemical gradient. I will spend little time here reviewing the function of
sea snake salt glands since Dunson (1975, 1976, 1979a, 1981a), Peaker
and Linzell (1975) and Minnich (1979) have already done so. The sea
snake salt gland excretes a hyperosmotic solution of sodium chloride
when there is a rise in osmotic pressure of the plasma. Although all
sea snakes possess salt glands, they are so small in some species that
their function is greatly reduced. Any possible link between such
reduced size and secretory capacity of salt glands and the functioning
of other parts of the osmoregulatory system remain obscure at present.
For example *Pelamis* has a large salt gland, but a skin relatively impermeable to water. Another large salt gland form, *Aipysurus laevis*,
has a much more permeable skin. Thus there is a definite need for comparative studies on the integrated functioning of the skin, salt gland
and kidney-cloaca-gut complex in selected species. It would also be
extremely valuable to know when and for how long the salt glands actually secrete after feeding. Do the tiny salt glands of certain *Hydrophis*
secrete continuously or do these species tolerate wider variations in

body fluid concentrations after a salt load is ingested ? What if the salt gland is removed; can fed snakes maintain water and salt balance without it ?

TABLE 2. Influx (I) and efflux (E) of water, sodium, and chloride across whole fresh snake skins (values in μmoles/cm^2.h, mean ± SD). From Dunson and Stokes (1983) and Stokes and Dunson (1982).

Species and Flux	Water	Sodium	Chloride
Marine :			
Pelamis platurus			
I	57+25[a]		
E	40+16[a]		
Aipysurus laevis			
I	238+17[a]	0.035+0.027[a]	0.003+0.003[a]
E	133+58[a]	0.105+0.086[a]	0.019+0.010[a]
Fresh water :			
Regina septemvittata[b]			
I	311+128[a]	0.461+0.364	0.378+0.174[a]
E	573+212[a]	0.229+0.238	0.028+0.027[a]

[a] I differs significantly from E ($\alpha < 0.06$) by Wilcoxon signed rank test.

[b] Shed skins; Br flux instead of Cl

Another interesting variable among sea snake salt glands is the concentration of the secreted fluid. There are different levels of secretion concentration; some species have approximately the same concentration as sea water and others are considerably higher (Dunson and Dunson, 1974). A remarkable feature is that the absolute concentration of the secretion has no relation to the overall rate of excretion. Variations in the sizes of the glands are related to secretory rate. Yet the biochemical or ultrastructural features which lead to elaboration of fluids quite different in concentration are not known. It would certainly be fruitful to examine tubule lengths as well as cellular ultrastructure to cast light on this question.

Recent detailed studies of renal function in *Aipysurus laevis* have provided a great deal of new information (Yokota et al., 1982; Benyajati et al., 1982). The kidneys of sea snakes show a few specializations for a marine existence, primarily a low glomerular filtration

rate and[a]net tubular secretion of magnesium. The glomerular filtration
rate of snakes loaded with sea water or a hyperosmotic sodium chloride
solution increased from 1.6 to 7.1 ml/kg.h. However the ureteral urine
was never more than slightly hyperosmotic to the plasma (maximum ratio
of about 1.2 at low urine flow rates). At urine flows above 1.0 ml/kg.
h, the osmotic urine/plasma ratio was constant (about 0.8), suggesting
little variation in the permeability of the distal nephron to water at
high urine flows. A very exciting finding was the demonstration, for
the first time in reptiles, of net water and sodium secretion by the
renal tubules. There were also increases in the fractional excretion
of sodium after a sea water load, indicating that further study of the
relative roles of the kidneys and the salt gland are warranted. The
possible role of cloacal urates in excretion of bound cations also
deserves attention (Minnich, 1979).

In summary it appears that the integument is the major route
of uptake and loss of water in fasting sea snakes. In some species water
exchange is quite rapid, although we are at a loss to explain exactly
what advantage this may have. The skin also serves as the single most
important organ in restricting intake of salts from the environment,
as a consequence of its semipermeable nature. Although the functioning
of the salt gland as the major route of excretion of sodium chloride
is clear, its overall quantitative role in osmoregulation remains
uncertain. This is a consequence of the extreme impermeability of the
integument to salts and the very slow net rate of water loss. Dehydration proceeds so slowly in fasting animals that the salt gland may not
often be stimulated to secrete. Long term sequential studies of individual snakes, exposed to periods of fasting and feeding would be very
helpful in quantifying the role of the salt gland. It would also be
especially interesting to examine the responses of dehydrated snakes
to simulated rainfall to see if they would drink the surface film of
brackish water. Dunson and Robinson (1976) showed that *Pelamis* would
drink fresh water in the laboratory. Since this species spends a great
deal of time floating at the surface, it is ideally situated to behaviorally osmoregulate by drinking rain water. The salt gland would then
be important primarily during dry seasons, or perhaps after ingestion
of prey.

VI. The Estuarine Specialists : *Cerberus, Malaclemys,* & *Crocodylus acutus*

The dog-faced water snake, *Cerberus*, is an animal well adapted
for long term survival in sea water. Yet as discussed above, it has

developed its osmoregulatory adaptations independently of the true sea
snakes (Hydrophiidae). The evidence consists of its presence in a
different family (Colubridae) and its unique permaxillary salt gland
(Dunson and Dunson, 1979). This tiny salt gland secretes at a very low
rate (about 15-20μ moles sodium/100 g body mass.h) and dedifferentiates
except when the snake is dehydrated (plasma above 150 mM sodium). Yet
it apparently provides a crucial backup source of sodium efflux when
other systems fail. *Cerberus* is low in integumentary permeability to
sodium. It has a very low overall body sodium influx and a low net
rate of water loss. Thus it dehydrates extremely slowly even when
fasting and readily rehydrates by drinking fresh water when it is
available. *Cerberus* normally shows no preference for habitat water
salinity, but a severely dehydrated snake did prefer fresh to sea
water (Zug and Dunson, 1979). Water exchange in fasting *Cerberus* in
sea water is almost entirely through the skin; vitually all sodium
influx is oral. The only major difference between the osmoregulatory
system of *Cerberus* and that of the true sea snakes may be acclimation
to low salinity conditions. An estuarine species may expect to perio-
dically encounter brackish water, whereas a pelagic snake may not,
except briefly on the surface after rainfalls. Thus the open sea
dwellers have little reason to develop a mechanism which de-activates
the salt gland. This question has not been well studied, but preliminary results for *Pelamis* (Dunson and Dunson, 1975) indicate that this sea
snake does not undergo a decrease in salt gland mass or Na-K ATPase
content after prolonged fresh water immersion. In contrast the premaxillary gland of *Cerberus* dedifferentiated after fresh water exposure
(Dunson and Dunson, 1979).

Another estuarine species, the diamondback terrapin (*Malaclemys*),
shows a very similar pattern of salt gland shut-down when plasma
sodium levels are low (Dunson and Dunson, 1975). This turtle has moved,
in an evolutionary sense, closer to a fully marine life style than
Cerberus, *Acrochordus*, or *Crocodylus acutus*. It does not occur in
inland fresh water habitats, although it can live in fresh water.
Instead it is found solely in coastal areas. As might be expected it
has a whole host of physiological adaptations for hypoosmoregulation,
including a few new features not seen in the snakes. For example
Malaclemys can expand the interstitial fluid volume to a large value
(19% body mass) when alternately exposed to a source of salts and
fresh water. Subsequently during dehydrating conditions this fluid
may be used as a source of free water by excreting the contained salts

with the salt gland. Another osmoregulatory strategy used by *Malaclemys* but not demonstrated in sea snakes is the accumulation of urea in the extracellular fluids to raise the osmotic pressure during dehydration (Gilles-Baillien, 1970). Dehydrated *Malaclemys* also develop very high potassium and low sodium levels in the bladder urine in concert with an increased Na-K ATPase content of the lachrymal salt gland (Robinson and Dunson, 1976). Despite these and other striking physiological adaptations (see reviews by Dunson, 1979a, and Minnich, 1979, 1982), *Malaclemys* also utilizes behavioral osmoregulation. When dehydrated it will seek out sources of fresh water, including rain running off a marsh. However *Malaclemys* is so impermeable to sodium and so resistant to dehydration that it often takes months for its plasma to increase in concentration when terrapins are placed in sea water. This poses an experimental problem since some investigators have naively assumed that acclimation occurs much more quickly. It is necessary to closely monitor plasma concentrations to determine if a terrapin immersed in a solution is actually undergoing internal changes.

Our third estuarine specialist, the American crocodile (*Crocodylus acutus*), is an interesting variation on the now familiar themes of skin permeability, salt glands, and fresh water drinking. In the southern Florida everglades the adults spend much of the year in fresh water swamps and marshes. The females move out into salt water areas of Florida Bay during the nesting season and the young are hatched in these saline regions. For some time there has been uncertainty as to the fate of these hatchlings since they fare poorly in the laboratory in sea water (Dunson, 1970, 1982; Evans and Ellis, 1977). Yet apparently they do survive in the wild in rather saline areas on northern Key Largo. A salt gland is present in *C.acutus*, but it has been stimulated to secrete only by mecholyl injection, not with salt loads (Taplin et al., 1982). Several investigators have also been unable to measure any significant increase in sodium efflux from salt loaded hatchlings (Dunson, 1970, 1982; Evans and Ellis, 1977; Mazzotti and Dunson, submitted). Young *C.acutus* have a very high rate of net mass loss (water loss) in sea water and in general seem poorly adapted physiologically for life in saline waters). Their major adaptations seem to be a low sodium influx, a high rate of growth to a size more resistant to high salinities, and behavioral osmoregulation by drinking of rain water (Mazzotti and Dunson, submitted). In marked contrast it is claimed that the Australian crocodile, *C.porosus*, is well adapted for life in salt water (Taplin, 1982). *C.porosus* does appear to have a greater capacity for extracloacal sodium excretion

than *C.acutus*; field data and limited laboratory tests demonstrate
growth in highly saline water (Grigg et al., 1980). However the studies
of *C.acutus* have focused on recently hatched animals, whereas very few
of this age class of *C.porosus* have been studied. Both species occur
in nature in both fresh and saline habitats in some areas. The joint
utilization of such chemically dissimilar habitats would seem to be
disadvantageous, since adaptations useful in one area would be unpro-
ductive in the other. It may be that such switching in habitat use is
only maintained in regions where there is a vacant niche for a large
aquatic carnivore (as in Caribbean islands lacking alligators or
caimans or in Australia where the other crocodilian differs in size
and the habitat undergoes a pronounced dry season change). Such an
explanation may also apply to the restriction of *Malaclemys* to the
estuary. Penetration of the fresher regions of rivers may be prevented
by "competitive" interactions with the abundant fresh water turtle
fauna of North America. These same fresh water species are excluded
from the estuary by a salinity barrier.

VII. The Mangrove Snake : An Incipient Estuarine Species

The North American salt water subspecies of the banded water
snake, *Nerodia* (previously *Natrix) fasciata*, are very close to becoming
separate species. They are generally isolated from the fresh water
parent subspecies (N.f.*pictiventris*) by their limitation to coastal
brackish and marine waters. Yet they do interbreed freely in disturbed
habitats. Although only subspecifically distinct, they differ morphologically, behaviorally, and physiologically from the fresh water
race (Dunson, 1980). This situation must represent an early stage by
which *Malaclemys* diverged from riverine *Graptemys*, gradually developing
more and more adaptations for coastal life. The *N.fasciata* subspecies
complex reveals with great clarity exactly what the minimal level of
adaptation is for survival in saline habitats. There is absolutely no
doubt that they are competent in completing their life cycle in high
salinity swamps and marshes. This is most clear in the Florida Keys
population of *N.f.compressicauda*, which live on isolated mangrove islands without a permanent source of fresh water. Most individuals
maintain plasma sodium levels near 150 mM, although occasional instances
of naturally dehydrated snakes are found (Dunson, 1980). In the laboratory this form and *N.f. clarki* survive for long periods even while
fasting in 35 ppt sea water. This is an important point since *C.acutus*
larger in size can not do the same, even when fed (Dunson, 1982;
Mazzotti and Dunson, submitted). The skin of *N.f.compressicauda* is

very low in permeability to sodium and water; the snakes thus have a
very low rate of net water loss and net sodium uptake in sea water.
The major question that remains is whether or not these snakes possess
a salt gland. *N.f. compressicauda* was one of the species originally
tested by Schmidt-Nielsen and Fänge (1958); they were unable to
stimulate any salt gland secretion with mecholyl. I have tried to
find a salt gland but have been unable to decide the issue. The problem is that extremely small salt glands are difficult to study.
In *Cerberus* the tiny premaxillary salt gland excretes sodium at about
15-20 μmoles/100 g body mass.h (Fig.1). In dehydrated snakes kept
dry this is evident as a salty deposit around the anterior end of the
head (Dunson and Dunson, 1979). Whole body sodium effluxes from a group
of eight wild-caught *N.f. compressicauda* injected with salt loads were
about 12μmoles/100g.h (Dunson, 1980). Three of the snakes had values
of 16-18 μmoles/100g.h. This indicates that a small salt gland might
be present since control effluxes were only 1-5 μmoles/100g.h. A
recent attempt to repeat this failed to achieve such an increased
rate of efflux after salt loading in three snakes from Florida Bay
(Dunson, unpub.obs.). Their sodium effluxes (2.6-3.7μmoles/100g.h)
did not change after injection of 3mmoles NaCl/100g. At the end of the
two month efflux period plasma sodium levels were elevated (211-243
mM). A careful ultrastructural examination of each of the head glands
is needed, as well as a partitioning of the sodium efflux to determine
how much is cephalic. If there is an exocrine gland excreting sodium
chloride in *N.f.compressicauda*, it must be the smallest of any known
reptilian salt gland.

The salt water *Nerodia* are thus a type case illustrating that,
for a fish eating reptile, very little extracloacal excretory capacity is needed for survival in sea water. This confirms what was suspected by the finding that some of the true sea snakes osmoregulate with
tiny salt glands with maximum excretory rates of about 25 μmoles sodium/
100g.h. In captivity, dehydrated *N.f. compressicauda* and *clarki* readily
drink fresh water and they most likely utilize ephemeral sources of
brackish water in the wild. A man-made pond of 14 ppt on Chokoloskee
Island, Florida, was teeming with *N.f.compressicauda*, while none were
seen in the nearby bay at 22ppt (Dunson, 1979b). In captivity this
form also shows a strong preference for fresh water over saline waters
(Zug and Dunson, 1979), although in nature it obviously does not distri-
bute itself in this fashion. Interbreeding and possible "competitive"
interactions with fresh water congenors may restrict this phenotype to
saline waters.

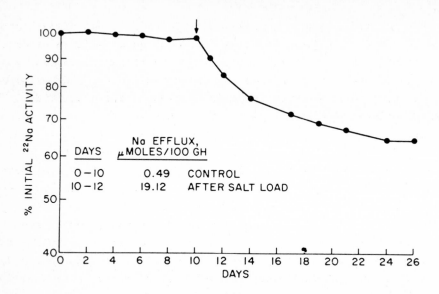

Fig.1. Radioactive sodium efflux in sea water of the homalopsine snake Cerberus rhynchops before and after injection of a NaCl load (2 mmoles/100g at arrow). The increase in efflux on days 10 to 12 is interpreted as the secretion of a suddenly activated premaxillary salt gland (from Dunson and Dunson, 1979).

Chelydra and *Kinosternon* : Invaders from the Fresh Water Fringe

Starting with the true sea snakes we have now examined a series of species showing a cline in the ability to cope with immersion in sea water. However all of these forms are able, in at least some part of their life cycle, to tolerate full strength sea water for long periods. These final two examples are truly fresh water turtles that show the first signs of development of estuarine adaptations. They are unusual in this regard since most fresh water reptiles die rapidly when placed in 35 ppt sea water. An often cited review by Niell (1958) erroneously claims that numerous species of reptiles utilize salt water habitats along the coast, especially in Florida. The problem with these locality data is that there are no salinity measurements reported nor any studies of laboratory tolerance. Most of these cases are simply accidental specimens or represent brackish estuaries below 10ppt which pose no osmoregulatory problem for fresh water reptiles.

The Key mud turtle, *Kinosternon b. baurii*, lives in small ponds on islands of the lower Florida Keys. Although experts disagree on its exact taxonomic status, it is at most weakly differentiated from mainland *K. baurii*. However it lives on small islands surrounded by salt water and its home ponds are subject to seasonal drying and wide changes in salinity. Dunson (1979c, 1981b) studied this turtle both in the field and the laboratory and found that it exists in seasonal brackish ponds primarily by means of behavioral osmoregulation. Turtles in this genus are generally known for their ability to travel overland and aestivate on land when ponds dry up. Thus *K. b. baurii* simply leaves a pond when the salinity reaches about 15 ppt (Fig.2). It is easy to imagine that a strong selective pressure would favor the development of any new traits in this population that would allow the turtles to remain in the ponds as the water becomes more saline. In addition there are many ponds and mangrove swamps that are too saline to support populations of mud turtles. These are potential habitat if an island race were to develop with greater osmoregulatory powers.

Two other chelonians, *Kinosternon subrubrum* and *Chelydra serpentina* are common inhabitants of coastal creeks and island ponds in the eastern U.S. They are considerably more tolerant of high salinity than other fresh water turtles and can colonize habitats unavailable to species of lesser tolerance (Dunson, unpub.obs.). Estuarine waters are often extremely rich in food and certain predatory niches may be unfilled because of the rigorous environment. This may explain the occurrence of *Chelydra* in considerable abundance in tidal creeks. We do not yet understand the extent of any physiological adaptations of these recent migrants into the estuary. *Chelydra* has a very low rate of sodium influx in sea water, but the rate of mass loss while fasting is higher than for estuarine specialists. Hatchling *Chelydra* also require considerably lower salinities for growth than do *Malaclemys* hatchlings. Further study of these incipient estuarine forms may lead to important insights into the earliest stages of the evolution of marine adaptations.

VIII. A New Model for Diffusion Across Reptilian Skin

Aside from the discovery of a new type of salt gland in crocodiles by Taplin and Grigg (1981), there is little new to add to previous reviews on the physiology of reptilian salt glands. In contrast there has been a considerable increase recently in our knowledge of reptilian skin. Thus it is worthwhile considering these new findings briefly and examining their implications for osmoregulation in marine reptiles. These data are divided into three areas, (a) the general role of lipids in controlling water and solute permeability, (b) the theory that reptilian skin has minute channels of at least two sizes that permit trans-integumental diffusion, and (c) the hypothesis that uni-directional diffusion rates across the integument (influx and efflux) may be specifically hindered for osmoregulatory reasons by some mechanism within the channels.

Roberts and Lillvwhite (1980) described a 15-fold increase in the rate of evaporative water loss across the shed skin of the rat snake, *Elaphe obsoleta*, after extraction of lipid. Partial denatura-tion of the skin protein led to only a two-fold increase. These measurements were made across an air-water interface with the shed skin separating the two phases. Similar results have been obtained on mammalian skin (see review by Lillywhite and Maderson, 1982). While such an experiment strongly implicates lipids as agents controlling water permeability, it offers little information on how they might be configured in the keratin membrane. This issue has however been addressed by Stokes and Dunson (1982a) who postulate a model to explain the much more rapid diffusion of water than sodium across snake skin. Stokes and Dunson (1982a) and Ljungman and Dunson (1983) found that the skin lipid extraction effect applied to water-water as well as to water-air interfaces. Since water diffuses much more rapidly than electrolytes, and is much smaller, it is possible that there are at least two major sizes of channels through the skin that permit diffusion. The water channels must be smaller and much more numerous than the ion channels. Stokes and Dunson (1982a) suggested that these channels are lined with lipids of specific properties for each species, and that removal of these by extraction enlarges the effective channel size and greatly increases the diffusion rate. This is a reasonable possibility but there is another which may be more plausible. Inter-cellular lipid lamellae extruded from the cells may form bilayers between the keratin protein framework (Wertz and Downing, 1982; Landmann, 1979). The rate limiting channels could then be located within these lipid bilayers. After lipid extraction, diffusion is limited by larger diameter keratin channels, perhaps located at the former cell junctions. Interspecific variation in permeability is then defined by the nature of the lipid bilayers. It is also possible that the integumentary lipid layers do not contain any pores or channels and that substances cross by a solubility-diffusion mechanism

(Finkelstein, 1976; MacDonald, 1976). A fruitful avenue for further research would be biochemical characterization of the skin lipids from species varying widely in permeability, including reconstitution in liposomes. Another route to follow would be to dissolve and reconstitute the keratin framework, with and without lipids present. Stokes and Dunson (1982b) have made some preliminary progress in this latter area , but the keratin films prepared were far more permeable than lipid extracted shed skins. Improved methods for reconstituting keratin sheets are needed.

An unusual property of snake skins first discovered by Stokes and Dunson (1982a) in fresh water snakes and later extended to marine species (Dunson and Stokes, 1983) is asymmetric diffusion. This unequal diffusion down a chemical gradient, depending on the direction of movement across the skin, obviously must be reflected in any model of the skin channels. The most striking characteristic of this polarization of water and ion fluxes is that each is asymmetric in opposite directions, in a way appropriate for osmoregulatory homeostasis (Table 2). We are obviously in no position yet to advance a firm hypothesis for the mechanism behind asymmetric diffusion in vitro, nor can we be certain that a similar process occurs in vivo without further study. Possible artifacts also need to be examined. However it is certainly possible that such an asymmetric effect could be achieved by differences in the state of hydration of hydrophobic and hydrophilic ends of lipid molecules surrounding the diffusion channels. A preferential swelling at one end could partially occlude the channel opening and restrict diffusion from one end only. Osmotic pressures and salt concentrations are different on the two sides of the skin under natural conditions. The degree of swelling could be related to such concentration differences applied in vitro with distilled water and 1 M sodium chloride as test solutions. These studies, while incomplete, suggest a completely new way of examining the contribution of the integument to osmoregulation of marine reptiles, and possibly of other vertebrates.

Acknowledgments

The continued financial support of the U.S. National Science Foundation is deeply appreciated. These studies are currently supported by NSF grant DEB BSR 8212623. J.Minnich provided critical comments on the manuscript.

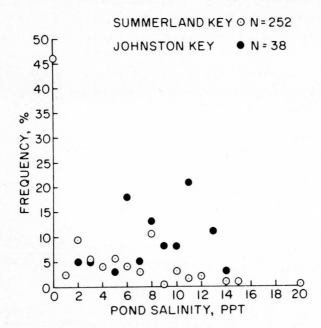

Fig.2. *The pond salinities at which mud turtles (Kinosternon b. baurii) were captured in the lower Florida Keys. All but one turtle were found at 15 ppt or below. This apparent preference for lower salinities was confirmed by tracking individual turtles during seasonal changes in pond water levels (from Dunson, 1981b).*

REFERENCES

Benyajati S. et al. (1982). Secretion and reabsorption of Mg and K by the kidney of the sea snake , Aipysurus laevis. Fed.Proc. 41(4) : 1005.

Dantzler W.H. (1976). Renal function (with special emphasis on nitrogen excretion). In : Biology of the Reptilia. Gans C. and Dawson W.R. (eds). Physiol.A. Vol.5 Acad.Press, New York. p.447-503.

Dunson M.K. and Dunson W.A. (1975). The relation between plasma Na concentration and salt gland Na-K ATPase content in the diamond-back terrapin and the yellow-bellied sea snake. J. Comp. Physiol. 101 : 89-97.

Dunson W.A. (1970). Some aspects of electrolyte and water balance in three estuarine reptiles, the diamondback terrapin, American and salt water crocodiles. Comp. Biochem. Physiol. 32 : 161-174.

Dunson W.A. (1975). Salt and water balance in sea snakes. In : The Biology of Sea Snakes. W.A.Dunson (ed.). Univ. Park Press, Baltimore, p. 329-353.

Dunson W.A. (1976). Salt glands in reptiles. In : Biology of the Reptilia. Gans C. and Dawson W.R. (eds). Physiol.A. Vol.5 Acad.Press, New York. p. 413-445.

Dunson W.A. (1978). Role of the skin in sodium and water exchange of aquatic snakes placed in sea water. Amer. J. Physiol. 235 : R151-159.

Dunson W.A. (1979a). Control mechanisms in reptiles. In : Mechanisms of Osmoregulation in Animals. R.Gilles (ed.). Wiley Interscience New York. p.273-322.

Dunson W.A. (1979b). Occurrence of partially striped forms of the mangrove snake *Nerodia fasciata compressicauda* Kennicott and comments on the status of *N.f. taeniata* Cope. Fla. Sci. 42(2) : 102-112.

Dunson W.A. (1979c). Salinity tolerance and osmoregulation of the Key mud turtle, *Kinosternon b.baurii*. Copeia 1979(3) : 548-552.

Dunson W.A. (1980). The relation of sodium and water balance to survival in sea water of estuarine and fresh-water races of the snakes *Nerodia fasciata*, *N. sipedon* and *N.valida*. Copeia 1980(2) : 268-80.

Dunson W.A. (1981a). Control of secretion in reptilian salt glands. In : Advances in Physiological Sciences. Vol.18. Environmental Sciences. F. Obal and G. Benedek (eds.). Proc. 28th Int. Cong. Physiol. Sci., Budapest. Pergamon Press, New York. p.31-41.

Dunson W.A. (1981b). Behavioral osmoregulation in the Key mud turtle, *Kinosternon b. baurii*. J.Herpetol. 15(2) : 163-173.

Dunson W.A. (1982). Salinity relations of crocodiles in Florida Bay. Copeia 1982(2) : 374-385.

Dunson W.A. and Dunson M.K. (1974). Interspecific differences in fluid concentration and secretion rate of sea snake salt glands. Amer. J. Physiol. 227 : 430-438.

Dunson W.A. and Dunson M.K. (1979). A possible new salt gland in a marine homalopsid snake (*Cerberus rhynchops*). Copeia 1979(4) : 661-672.

Dunson W.A. and Robinson G.D. (1976). Sea snake skin : permeable to water but not to sodium. J. Comp. Physiol. 108 : 303-311.

Dunson W.A. and Stokes G.D. (1983). Asymmetrical diffusion of sodium and water through the skin of sea snakes. Physiol. Zool. 56(1): 106-111.

Evans D.H. and Ellis T.M.(1977). Sodium balance in the hatchling
American crocodile, *Crocodylus acutus*. Comp. Biochem. Physiol.
58A : 159-162

Finkelstein A.(1976). Water and nonelectrolyte permeability of lipid
bilayer membranes. J. Gen. Physiol. 68 : 127-135.

Gilles-Baillien M. (1970). Urea and osmoregulation in the diamondback
terrapin *Malaclemys centrata centrata* (Latreille). J. Exp.
Biol. 52 : 691-697.

Grigg G.C. et al. (1980). Survival and growth of hatchling
Crocodylus porosus in saltwater without access to fresh
drinking water. Oecologia 47 : 264-266.

Landmann L.(1979). Keratin formation and barrier mechanisms in the
epidermis of *Natrix natrix* (Reptilia : Serpentes) : an ultrastructural study. J. Morph. 162 : 93-126.

Lillywhite H.B. and Maderson P.F.A. (1982). Skin structure and permeability. In : Biology of the Reptilia. Gans C. and Pough F.
(eds.) Physiol. C. Vol.12 Acad.Press, New York. p.391-442.

Ljungman T.N. and Dunson W.A. (1983). Integumentary water and sodium
permeability of the yellow anaconda, *Eunectes notaeus*.
Comp. Biochem. Physiol. 76A (1) : 51-53.

MacDonald R.C. (1976). Energetics of permeation of thin lipid
membranes by ions. Biochim. Biophys. Acta 448 : 193-198.

Mazzotti F. and Dunson W.A. Submitted. Adaptations of *Crocodylus
acutus* for life in saline water.

Minnich J.E. (1979). Reptiles. In : Comparative Physiology of Osmoregulation in Animals. Maloiy G.M.O (ed.) Vol.1. Acad.Press,
New York. p.391-641.

Minnich J.E. (1982). The use of water. In : Biology of the Reptilia.
Gans C. and Pough F. (eds). Physiol. C. vol.12, Acad.Press
New York. p.325-395.

Neill W.T. (1958). The occurrence of amphibians and reptiles in
salt water areas; and a bibliography. Bull. Mar. Sci. Gulf
Carib. 8 : 1-97.

Peaker M. and Linzell, J.L. (1975). Salt glands in birds and reptiles.
Cambridge Univ.Press. Cambridge.

Roberts J.B. and Lillywhite H.B.(1980). Lipid barrier to water
exchange in reptile epidermis. Science 207 : 1077-79.

Robinson G.D. and Dunson W.A. (1976). Water and sodium balance in
the estuarine diamondback terrapin (*Malaclemys*). J. Comp.
Physiol. 105 : 129-152.

Schmidt-Nielsen K. and Fänge R. (1958). Salt glands in marine reptiles. Nature (Lond.) 182 : 783-785.

Stokes G.D. and Dunson W.A. (1982a). Permeability and channel structure of reptilian skin. Amer. J. Physiol. 242 : F681-689.

Stokes G.D. and Dunson W.A. (1982b). Passage of water and electrolytes through natural and artificial keratin membranes. Desalination 42 : 321-328.

Taplin L.E. (1982). Osmoregulation in the estuarine crocodile, *Crocodylus porosus*. PhDthesis Univ. of Sydney, Sydney, Australia.

Taplin L.E. and Grigg, G.C. (1981). Salt glands in the tongue of the estuarine crocodile, *Crocodylus porosus*. Science 212 : 1045-1047.

Taplin L.E. et al. (1982). Lingual salt glands in *Crocodylus acutus* and *C.johnstoni* and their absence from *Alligator mississipiensis* and *Caiman crocodilus*. J. Comp. Physiol. 149 : 43-47.

Wertz P.W. and Downing D.T. (1982). Glycolipids in mammalian epidermis: structure and function in the water barrier. Science 217 : 1261-2.

Yokota S.D. et al. (1982). Renal function in the sea snake, *Aipysurus laevis*. Fed. Proc. 41(4): 1005.

Zug D.A. and Dunson W.A. (1979). Salinity preference in fresh water and estuarine snakes (*Nerodia sipedon* and *N.fasciata*).Fla. Sci. 42 : 1-8.

Regulation of NaCl and water absorption in duck intestine

E.SKADHAUGE, B.G.MUNCK and G.E.RICE

I. INTRODUCTION

Post-renal absorption of NaCl and water plays an important role in
osmoregulation of terrestrial, seed-eating birds, especially during dehydration and NaCl depletion (see Skadhauge 1973, p. 53, Skadhauge
1981, p. 100-112). NaCl-loading reduces cloacal absorption of salt
and water, as observed both in vivo (Rice & Skadhauge 1982a, Skadhauge
1967, Thomas & Skadhauge 1979) and in vitro in the domestic fowl (Choshniak et al. 1977, Lind et al. 1980, Thomas et al. 1980).

An interesting problem is therefore what happens in marine birds
with functional nasal salt glands when they receive a high NaCl intake?
Will absorption of NaCl and water in the cloaca remain high? This would
allow urinary water to be recuperated as "free water" by the elimination via the salt glands of cloacally resorbed NaCl. This sequence of
events was suggested by Schmidt-Nielsen et al. (1963), but the crucial
role of the cloaca in the hypothesis has never been examined.

In this paper we report measurements, in vivo and in vitro, of ion
and water transport in the cloaca of domestic ducks fed a diet with
either low- or high-NaCl content. The findings give quantitative proof
for the hypothesis of Schmidt-Nielsen et al. (1963).

Since Crocker & Holmes (1971) have reported a remarkable change of
ion and water absorption in the small intestine as functions of the Na-
Cl content of the diet, we also measured the absorption in ileum.

II. MATERIAL AND METHODS

Animals. Adult domestic Pekin ducks (mean BWT = 3.0 kg, n = 25)
were maintained on high-NaCl diet with either 0.5 % (in vitro studies)

or 1.0 % (in vivo studies) NaCl-drinking water, or low-NaCl diet and
tap water, for fourteen days prior to the measurement of transport parameters (see Thomas & Skadhauge 1979 for diet composition).

In vitro experiments. Blood was drawn by heart puncture into hepa-
rinized syringes; the birds were decapitated and the lower intestine
removed and rinsed in a Krebs-phosphate medium (Choshniak et al. 1977).
The serosal surface area of coprodeum and colon was determined by a
ruler. The coprodeal mucosa was stripped by free dissection and the colon mucosa by scraping with a razor blade. The isolated mucosal tissues were mounted in Ussing-chambers at $38^{o}C$. The tissues were kept
at zero electrical potential difference (PD) by short-circuiting with
an automatic device with compensation for bath resistance. Short-circuit current (SCC) was monitored continuously, and the tissues were
unclamped for 24 s each 5 min for measurements of PD. Two or three
tissues were mounted for each epithelium, and stable SCC and PD were
recorded before and after addition of 15 mM glucose and 4 mM leucine
and lysine. Presence of glucose is necessary for coprodeal function
in the hen (Choshniak et al. 1977), and in concert with glucose the
amino acids fully stimulate the electrolyte transport in chicken colon (Lind et al. 1980).

In vivo experiments. Anaesthesia and surgery were as described by
Rice and Skadhauge (1982b). The double recirculating, closed-perfu-
sion technique and sampling method described by Rice and Skadhauge
(1982b) were used in all experiments. It permits the separate, simultaneous intra-luminal perfusion of colon and small intestine (ileum). The segment of small intestine perfused was defined as the first
15 cm anterior to the termination of the caeca. The caeca run retro-
grade and are closely associated with ileum.

Two perfusion solutions were used: first the Krebs-phosphate medium,
second this solution with addition of 15 mM glucose and 5 mM l-leucine
and l-lysine. The concentrations of the solutions after initial 10
min mixing period are presented in Table 1. Two 45-60 minute perfusion samples per segment were collected for each perfusion solution.
Between subsequent perfusion periods the segments were rinsed with 100
ml of body-temperature, iso-osmotic saline and flushed with air. Transmural PD was measured during each perfusion period, for colon and small
intestine, using the method of Thomas and Skadhauge (1979). At the
conclusion of the final perfusion period a blood sample was collected
via a brachial vein catheter and the animal was killed by intravenous
injection of 3 M KCl. The surface area of the segments were determined
by planimetry.

TABLE 1. Composition of in vivo perfusion fluids after 10 min perfusion (other constituents, see text).

	Osmolality mOsm/kg H_2O	Na mequiv/l	Cl mequiv/l	K mequiv/l	No. of exp.
Iso-osmotic perfusion fluid	309±0.6	155±0.7	160±0.5	9.0±0.3	35
Glucose-amino acid perfusion fluid	295±2.0	138±1.5	145±1.2	9.1±0.5	32

Analyses. Blood was centrifuged immediately after collection and the plasma osmolality and electrolyte concentrations (Na, K, and Cl) determined. Aliquots of plasma were frozen from the in vitro experiments and transferred on dry ice to Montpellier for analysis of aldosterone and corticosterone (courtesy of Dr. M. Jallagéas). On in vivo perfusion samples osmolality and concentrations of Na, K, and Cl, and ^{14}C-PEG-4000 were analysed. Osmolality was determined by freezing point depression (Advanced osmometer). Na and K were measured by internal standard flame photometry (Radiometer). Cl by citration (Radiometer) and ^{14}C-PEG in the Packard Tricarb Spectrometer. ^{14}C-counting was as described by Rice and Skadhauge (1982b). Corticosterone and aldosterone were analysed as described by Thomas et al.(1980).

Intestinal transport rates for water and electrolytes were calculated by the method of Skadhauge (1967). Absorption rates and SCC's were denoted positive in the mucosa-serosa direction (m-s), negative in the s-m direction. Results are reported as means ± standard error of the mean, and statistical significance determined by "Student"'s t-test.

III. RESULTS

Control experiments. Tissues from eight birds were used for in vitro experiments on each diet; four birds were investigated in vivo on low-NaCl diet, and five on high-NaCl diet. The plasma osmolality and electrolyte concentrations of all birds are reported in Table 2. The weight of dissected salt glands, and the plasma adreno-cortical steroid concentrations are reported in Table 3.

The serosal surface area of coprodeum and colon, measured in the in vitro experiments, averaged 6.0±0.4 cm^2 and 31.7±1.6 cm^2, respecti-

TABLE 2. Plasma osmolality and electrolyte concentrations.

Diet	Osmolality mOsm/kg H_2O	Na mequiv/l	Cl mequiv/l	K mequiv/l	No. of exp.
Low-NaCl	293±5	138±5	112±8	4.5±0.4	12
High-NaCl + 0.5 % saline	304±10	145±5	117±5	4.0±0.1	8
High-NaCl + 1.0 % saline	314±6	151±1	118±1	3.9±0.5	5
	$P<0.05$[x]	$P<0.05$[x]	NS[x]	NS[x]	

NS = non-significant.

x refers to comparison between "low" and "high" + 1.0 % saline.

TABLE 3. Weight of nasal salt glands and concentrations in plasma of aldosterone and corticosterone.

Diet	Weight of nasal salt glands mg/gland	Plasma aldosterone ng/dl	Plasma corticosterone µg/dl
Low-NaCl	157±16 (12) $P<0.001$	80±9 (8) $P<0.001$	2.8±0.4 (8) NS
High-NaCl + 0.5 % saline	392±27 (8) NS	15±3 (8)	1.8±0.3 (8)
High-NaCl + 1.0 % saline	438±13 (5)	-	-

vely. In the in vivo experiments colonic surface area was 36.4±1.8 cm^2 or 12.6±0.7 cm^2/kg. The area of the perfused ileal segments ave-raged 30.4±2.6 cm^2, or 10.8±1.4 cm^2/kg.

A. In vitro experiments

Coprodeum. Both SCC and PD were small, and not significantly chan-ged by the NaCl intake: average SCC and PD were 11±3 $\mu A/cm^2$ and 8.5± 1.6 mV (n = 14) on low-NaCl diet, and 12±3 $\mu A/cm^2$ and 7.0±1.4 mV (n = 14) on high-NaCl diet. The PD's were lumen negative. From these values an approximate resistance of 700 Ω cm^2 may be calculated. In no case was a stimulation with glucose or amino acids observed. Amiloride (10^{-4}M) reduced both the PD and the SCC to unmeasurable values.

Colon. The SCC was high on both high- and low-NaCl diets, and the tissues responded to glucose and amino acids but the responses were variable. The tissues from birds on low-NaCl diet had initial SCC of 66±8 μA/cm^2 and PD of 5.2±0.4 mV (n = 16), lumen-negative. Five of these tissues reacted to glucose and amino acids with an increase in SCC of 76±21 μA/cm^2 and in PD of 5.0±2.0 mV. These five preparations reacted to amiloride with a reduction of SCC of only 40±13 % whereas in the 11 preparations which did not respond to the non-electrolytes, SCC was reduced by 101±3 %. The resistance, with this diet, was around 75 Ω cm^2. The tissues from high-NaCl diet birds had an average SCC of 27±17 μA/cm^2 (n = 16). PD was 3.0±0.9 mV (n = 16), lumen-negative. After addition of glucose the SCC increased by 129±15 μA/cm^2, and with the subsequent addition of amino acids the total increase was 166±19 μA/cm^2 (n = 12). The increase in PD from the initial value was 15±3 mV. The calculated resistance was 100 Ω cm^2. Amiloride was without effect on SCC and PD.

B. In vivo experiments

Colon. Colonic transmural water flux (J_v) and net rates of electrolyte transport (J_{Na}, J_K, J_{Cl}) with the two perfusion solutions are presented in Table 4. With adaptation to low-NaCl diet colonic NaCl absorption and K secretion are high and associated with net Cl absorption. The solute-linked water flow averaged 243±77 μl/kg hr. The inclusion of glucose (15 mM) and amino acids (5 mM) in the perfusate did not result in any significant change of net electrolyte and water transport rates or transmural PD.

In ducks maintained on a high-NaCl diet NaCl absorption and K secretion were significantly lower (P<0.01) than values observed for colon from low-NaCl diet-adapted ducks. A pronounced glucose and amino acid stimulation of NaCl and water absorption was evident. Net NaCl absorption increased 2-3 fold and water absorption approximately 5 fold (Table 4). The apparent Na concentration of the absorbate was 172 mequiv/ l. The PD was increased by 66 %. No significant effect on K secretion was observed.

Small intestine. The transmural water and electrolyte transport rates in ileum (15 cm) are presented in Table 5. Transport rates in birds on a low-NaCl diet were approximately double the colonic trans-

TABLE 4. Duck colon: Transport rates for water and electrolytes.

Perfusion solution	J_V (µl/kg h)	J_{Na}	J_{Cl} (µequiv/kg h)	J_K	PD (mV)
	Low-NaCl diet				
Iso-osmotic (8)	243±77 NS	176±27 NS	92±16 NS	-66±11 NS	-23.6±7.0 NS
Glucose + amino acids (8)	513±133	202±31	121±25	-67±8	-25.8±5.1
	High-NaCl diet				
Iso-osmotic (10)	119±74 P<0.05	47±18 P<0.05	27±8 P<0.01	-31±7 NS	-23.6±5.7 P<0.05
Glucose + amino acids (10)	755±225	133±35	115±36	-47±2	-39.1±5.1

TABLE 5. Duck small intestine. Transport rates for water and electrolytes.

Perfusion solution	J_V (µl/kg h)	J_{Na}	J_{Cl} (µequiv/kg h)	J_K	PD (mV)
	Low-NaCl diet				
Iso-osmotic (8)	570±149 NS	247±34 NS	216±59 NS	-60±7 P<0.05	-1.0±5.4 NS
Glucose + amino acids (8)	875±242	211±48	158±56	-87±7	-6.3±1.9
	High-NaCl diet				
Iso-osmotic (10)	705±83 NS	159±16 P<0.05	155±20 P<0.05	-27±4.3 NS	2.3±2.0 P<0.05
Glucose + amino acids (10)	908±195	171±5	139±45	-58±13	-3.0±0.7

NS = non-significant.

port rates and were unaffected by glucose and amino acids. PD was always close to zero. With adaptation to high-NaCl diet, NaCl absorption and K secretion were significantly reduced (P<0.01), when compared to control values on low-NaCl diet. Glucose and amino acids re-

sulted in a small increase in Na absorption, K secretion, and PD (P < 0.05).

IV. <u>DISCUSSION</u>

 A.<u>Control data</u>. The osmotic stimulus induced by high-NaCl diet with 0.5 % (or 1.0 %) saline, was very conspicuous as the weight of the salt glands was augmented by 150 % (180 %). The weights were 2.6 % (3.0 %) of body weight. The average increase in plasma osmolality was 3.8 % (7.2 %). Similar increases in NaCl concentration were observed. As to be expected high-NaCl diet reduced plasma aldosterone by approximately 80 %. The osmotic stimulus used in this study was, however, not maximal as a 2 % saline drinking solution has been reported to increase plasma osmolality by 13 % (Deutsch et al. 1979).

 B.<u>In vitro experiments</u>. In coprodeum the SCC's were small. •It can by analogy with the domestic fowl be concluded that the net rate of ion absorption, presumably of Na, is small: the transport is not regulated as a function of the NaCl intake. When the small area is considered the net cation transport in coprodeum will be 86 times smaller than in colon. The coprodeum transport is therefore of little significance compared to that of colon. Coprodeum was accordingly not investigated <u>in</u> <u>vivo</u>. The net absorption of cation was on the low-NaCl diet only 4 % of that measured in the domestic fowl (Choshniak et al. 1977) and the resistance was the double.

 In colon the SCC's were high as in the domestic fowl (Lind et al. 1980) and glucose and amino acid stimulation was observed. This stimulation was associated with insensitivity to amiloride. The change from low- to high-NaCl diet therefore seems to cause a switch in the luminal membrane of the epithelial cells from a Na-channel to a non-electrolyte sodium coupled mechanism (Lind et al. 1980).

 C.<u>In vivo perfusion experiments</u>. A great advantage of <u>in</u> <u>vivo</u> perfusion is that this technique allows fairly precise measurements of net water absorption. This study showed a much higher rate of water absorption in colon (Table 4) than observed in the domestic fowl (Rice & Skadhauge 1982b). In the absence of non-electrolyte stimulation water absorption was higher on low-NaCl diet, but glucose and amino acids caused a significant (P<0.05) increase in Na, Cl and water absorption, with a nearly plasma-isoosmotic proportion to NaCl. This stimulation is considered physiological (i.e. normally occurring) as the chyme entering colon contains these non-electrolytes (see Skadhauge 1981, p. 33). <u>Figure 1</u> illustrates the rates of Na absorption

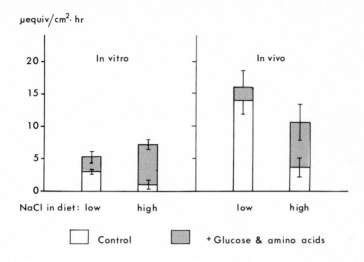

DUCK COLON: Net cation/sodium absorption

FIGURE 1. Net rates of cation/sodium absorption in colon.
Transport rates are reported for birds on high- and low-NaCl diets.
The controls are bathed in a Krebs-phosphate medium; stimulations with
glucose (15 mM) and amino acids (leucine and lysine, 5 mM) is shown.
Short-circuit currents were converted to transport of a monovalent cation. The stimulation is larger in colon from birds on the high-NaCl
diet.

in vitro and in vivo, and the effects of non-electrolytes. The SCC is
converted to transport of a monovalent cation.

The ileum had, as to be expected, high rates of NaCl and water absorption. The absorption rates were not consistently affected by the
addition of non-electrolytes. The transport rates calculated per cm^2
serosal surface area were not different from those of colon. The increased ileal absorption observed immediately after NaCl-loading (Crocker & Holmes 1971) was thus not confirmed. The present study was,
however, conducted on birds which had received the diets for several
days.
D. Reno-intestinal salt-gland recycling of NaCl. This paper clearly
shows that resorption of NaCl and water may proceed in the lower intestine of a bird with a functional nasal salt gland even on a high-
NaCl diet. The hypothesis proposed by Schmidt-Nielsen et al. (1963)
suggests that a fraction of ureterally excreted NaCl and water is resorbed in the lower gut, and the resorbed NaCl subsequently excreted,
with little water, through the nasal glands. This mechanism al-

SODIUM AND WATER TRANSPORT IN THE DUCK

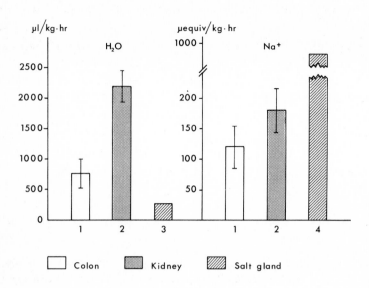

FIGURE 2.   Comparison of transport rates of sodium and water in colon,
kidney, and salt gland.
1. Rates of colonic absorption of water and Na (present study).
2. Rates of ureteral excretion of water and Na in ducks receiving 1.7
   % NaCl as drinking fluid (from Holmes et al. 1968).
3. The amount of water which must be let out through the salt gland
   (Na concentration 500 mequiv/l) in order to excrete the Na absorbed in colon (present study).
4. The high excretory capacity for Na of the salt gland after intravenous NaCl-loading is shown (from Lanthier & Sandor 1967).
   The difference between the first (1) and the third column (3) in
the left half of the figure demonstrate the water-conserving role of
the colon/salt gland interaction.

lows continued flushing of the insoluble uric acid to the cloaca, and
excretion of NaCl, but conservation of an important fraction of the
renally excreted water. In Figure 2 the rates of Na and water absorption in colon of the high-NaCl diet ducks are compared to renal excretion rates (Holmes et al. 1968). Approximately 1/3 of renally excreted water and 2/3 of Na can be resorbed. The colonic Na-absorption
is small compared to the secretory capacity of the salt gland, which
is around 1000 µequiv/kg hr (Lanthier & Sandor 1967). The amount of
water which will be lost through the salt gland with the excretion of

an amount of Na equivalent to that resorbed in the colon, was calculated (a Na concentration of the salt gland fluid of 500 mequiv/l was assumed). Figure 2 shows that approximately 500 µl water per kg hr is
retained by the colon-salt-gland interaction. As the intestinal area
for resorption includes not only colon (and coprodeum) but also the
caeca, the total resorption of water may be higher.

ACKNOWLEDGEMENTS

We thank Dr. M. Jallagéas, Laboratoire de Neuroendocrinologie, Université de Montpellier II, France, for the hormone analyses, and NOVO's
Fond for support. G.E. Rice received a CSIRO postdoctoral fellowship.

REFERENCES

Choshniak, I, Munck, BG and Skadhauge, E (1977) Sodium chloride transport across the chicken coprodeum. Basic characteristics and dependence on the sodium chloride intake. J Physiol (Lond) 271: 489-504

Crocker, AD and Holmes, WN (1971) Intestinal absorption in ducklings
(Anas platyrhynchos) maintained on fresh water and hypertonic saline.
Comp Biochem Physiol 40A: 203-211

Deutsch, H, Hammel, HT, Simon, E and Simon-Oppermann, C (1979) Osmolality and volume factors in salt gland control of Pekin ducks after
adaptation to chronic salt loading. J Comp Physiol 129: 301-308

Holmes, WN, Fletcher, GL and Stewart, DJ (1968) The patterns of renal
electrolyte excretion in duck (Anas platyrhynchos) maintained on freshwater and on hypertonic saline. J Exp Biol 48: 487-508

Lanthier, A and Sandor, T (1967) Control of the salt-secreting gland
of the duck. I. Osmotic regulation. Canad J Physiol Pharmacol 45:
925-936

Lind, J, Munck, BG, Olsen, O and Skadhauge, E (1980) Effects of sugars,
amino acids and inhibitors on electrolyte transport across hen colon
at different sodium chloride intakes. J Physiol (Lond) 305: 315-325

Rice, GE and Skadhauge, E (1982a) Colonic and coprodeal transepithelial
transport parameters in NaCl-loaded domestic fowl. J Comp Physiol B
147: 65-69

Rice, GE and Skadhauge, E (1982b) The in vivo dissociation of colonic and coprodeal transepithelial transport in NaCl depleted fowl. J. Comp Physiol B 146: 51-56

Schmidt-Nielsen, K, Borut, A, Lee, P and Crawford, E (1963) Nasal salt excretion and the possible function of the cloaca in water conservation. Science 142: 1300-1301

Skadhauge, E (1967) In vivo perfusion studies of the water and electrolyte resorption in the cloaca of the fowl (Gallus domesticus). Comp Biochem Physiol 23: 483-501

Skadhauge, E (1973) Renal and cloacal salt and water transport in the fowl (Gallus domesticus). Dan Med Bull 20, suppl 1: 1-82

Skadhauge, E (1981) Osmoregulation in birds. In: Farner, DS (ed.) Zoophysiology Vol. 12. Springer-Verlag, Berlin, 203 pp.

Thomas, DH and Skadhauge, E (1979) Dietary Na^+ effects on transepithelial transport of NaCl in hen (Gallus domesticus) lower intestine (colon and coprodeum) perfused luminally in vivo. Pflügers Arch 379: 229-236

Thomas, DH, Jallagéas, M, Munck, BG and Skadhauge, E (1980) Aldosterone effects on electrolyte transport of the lower intestine (coprodeum and colon) of the fowl (Gallus domesticus) in vitro. Gen Comp Endocronol 40: 44-51

PART II

Biophysical and Biochemical Aspects of "salt-transporting tissues" Studies

Cellular energy metabolism and its regulation in gills of fish and crustacea

C. LERAY

The ion osmoregulatory problems faced by aquatic animals have been extensively studied (Krogh 1939, Maloiy 1977) and the physiological mechanisms involved at the level of specialized interfaces are progressively well understood (Ussing & Thorn 1973, Gupta et al. 1977, Giebisch 1978, Gilles 1979a). Few studies have been devoted to energy expended for ion regulation in euryhaline crustacea and fish and the published results are conflicting. Some of these studies have been experimental (cf. Ref. in Leray et al. 1981), others have been theoretical (Potts et al. 1973, Fletcher 1975).

Several attempts have been made to correlate ion transport rates with oxygen consumption or other energetic aspects in various vertebrates tissues since the pioneer work of Zerahn (1956) but only one study was recently devoted to a fish excretory organ (Silva et al. 1980). The emphasis of much of these researches has been toward determining the coupling ratio between ions transported and oxygen molecules utilized in a variety of epithelial tissues (Mandel & Balaban 1981). Thus, much evidence was presented for the existence of a tight coupling mechanism between Na,K-ATPase-mediated ion transport and oxidative metabolism. According to the Whittam model (Blond & Whittam 1980) this coupling implicates the cellular adenylate pool both for energy provision to the ion pump and for respiration control at the mitochondrial level. In this model, schematically included in Fig.1, any increase in active ion transport would cause an increase in the rate of conversion of ATP to ADP in direct proportion to the energy requirements of the increased cellular activity. In turn, the ADP increase would serve as a feedback signal resulting in an acceleration of the oxygen consumption rate. Due to several technical limitations, several attempts to correlate directly nucleotide levels with function have failed using amphibian or mammalian salt transporting organs (Mandel & Balaban 1981). However, a

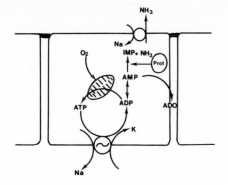

Fig. 1. Schematic representation of mechanisms implicating adenylate compounds in ion transport across epithelial cells. Prot. : membrane protease modifying AMP deaminase ; ADO : adenosine.

recent approach based on a combination of optical and biochemical methods (Balaban et al. 1980) applied to isolated rabbit cortical tubules provided experimental evidence in favour of the Whittam model. Since fish (Kirschner 1977, Potts 1977) and crustacea wood 1977) gills are the prime regions of ion transport at the body surface, they afford an opportunity to examine the relationship between the adenylate pool composition and changes in active ion transport either in steady or non steady state.

I. GILL ENERGETICS IN FISH

The ability of euryhaline fish to survive in waters of various salt contents and the various osmoregulatory efficiencies presented by different species were used in connection with chromatographic determination of gill nucleotides and if necessary with the determination of some physiological parameters. For this comparative study fish were transferred abruptly from fresh water to seawater. To prevent any possible anoxia effect on nucleotide levels, cellular metabolism was quenched in few seconds by throwing fish rapidly in liquid nitrogen. Extraction and HPLC analysis were processed by rapid and efficient methods (Leray et al. 1981). Due to the presence of an enclosed amount of blood in gill filaments, the true nucleotide content of the gill tissue must be calculated from the total nucleotide content, the blood nucleotide content and the blood content. Frozen blood was easily sampled in sinus venosus and atrium.

Rainbow trout studies : Upon transfer of rainbow trout (*Salmo gairdneri*) into sea water, we were able to describe (Leray et al. 1981) an initial "crisis" lasting about 30 h, characterized by large physiological changes following the abrupt increase in environmental salinity. After this first phase, which can lead to death in a varying percentage of animals, a stabilization period is observed during which the regulatory processes overcome the cellular dehydration and electrolyte loads.

The time course modifications of gill nucleotides is not synchronized with these general events since gill ATP shows an immediate and large fall 15 min after transfer to sea water leading to a 3 days steady level ; this level approximates again the freshwater value 3 days

later (fig. 2). The whole adenylate pool follows a similar trend during
3 days after the transfer but the decreased level is maintained till
the end of the experiment. These biochemical changes are consistent with
an immediate energy requirement of ion regulation leading to an increased
ATP utilization through the Na pump. The important fall in ATP indicates
that the rate of its utilization exceeds the rate of its synthesis until
a minimal value amounting to the half of the freshwater control value.
It can be assumed that gill ion transport is unaffected until this ATP
threshold is reached but for lower ATP content the ion extrusion can be
reduced and sea water survival compromised. Such a resistance of the
ion-transporting mechanisms to a fall in the supply of metabolic energy
was described in the rat liver cells (Van Rossum 1972). The fall and
the low steady level of the gill adenine nucleotides soon after the sea
water transfer likely reflect the inefficiency of their synthesis from
endogenous precursors in accordance with results on the enzyme equipment (Leray et al. 1979) and purine bases incorporations (Leray, unpublished data). The participation of the adenylate catabolism to ion
transport processes and the importance of the pathways leading to the
conversion of AMP into IMP, NH3 and adenosine must be emphasized (fig. 1).
First, the role of AMP deaminase which is abundant in trout gill epithelium (Raffin & Leray 1980) is well known in controlling the adenylate energy charge. The recent demonstration of its regulation by a membrane protease during sea water adaptation (Raffin 1983) throws new
lights on this metabolic step. Secondly, the dephosphorylation of AMP
into adenosine which is a potent vaso-active substance at the trout gill
level (Colin et al. 1979) would be of value in regulating the local blood
flow to support increased ion transport. All these transformations must
contribute to the adenylate pool decrease during the observed acceleration of ATP turnover. The ATP/ADP ratio has the lowest value at 4 and
72 h but has a similar value in freshwater and after 31 h in sea water
(fig. 3). After 6 days in sea water, this parameter has a higher value
than in control fresh water fish, but later, it approximates the control
value. If the 4 h shock likely results from the immediate oral salt
load leading to a steep rise of plasma electrolyte content, the 72 h
response concomitant with a drop in circulating ions concentration likely results from multiplication of chloride cells and/or Na,K-ATPase
sites. During the 10-day experiment, the ATP/ADP ratio and energy charge
of gill tissue display similar trends. Since the adenylate energy charge
is considered a key parameter controlling the balance between catabolic
and anabolic pathways (Atkinson 1977), the metabolic activity of the
gill tissue would fluctuate within a narrow specific range despite the

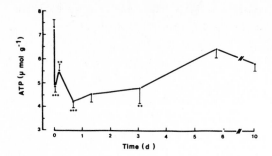

Fig. 2. *Gill ATP concentration in rainbow trout transferred from freshwater to sea water.*

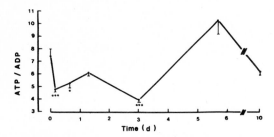

Fig. 3. *Gill ATP/ADP ratio in rainbow trout transferred from freshwater to sea water.*

high variations of adenylate compounds during the early "crisis" of sea-water adaptation. The unexpected return to a high energetic state in the gill after the preadaptative phase attested by a reduction of O_2 consumption together with elevated energy charge and ATP/ADP ratio can only be correlated with a reduced energy demand. Similar situations have been previously reported in euryhaline fish (Holmes & Stott 1960, Pequignot & Serfaty 1965, Stagg & Shuttleworth 1982) and can be explained, as suggested by Stuenkel & Hillyard (1980) and Stagg & Shuttleworth (1982), by an identical hydromineral stress in freshwater and sea water environments with comparable energy input at the gill level to support hypo- or hyper-regulation of ionic body composition. This conclusion is corroborated by the works of Rao (1968) and Farmer & Beamish (1969) on rainbow trout and *Tilapia* respectively which suggest that the magnitude of the osmotic gradient is more important that its direction in the energy requirements of osmoregulation. Furthermore, several studies have been carried out on gill Na^+,K^+-ATPase (Rev. in Stagg and Shuttleworth 1982) and their results suggest that its lowest activity occurs when ion active transports are the lowest, e.i., in isosmotic conditions. *Eel studies* : Similar investigations were performed on freshwater yellow eel (*Anguilla anguilla*) which is known to withstand rapid changes in the salinity of the external medium and also to be a better osmoregulator than rainbow trout. Thus upon transfer in sea water the salt load imposed by drunk water and passive diffusion is excreted by the gill at an increasing rate during 2 days thereafter a steady state is reached (Mayer & Nibelle 1970). In animal of similar size, gill Na^+,K^+-ATPase activity was seen to increase slowly during the first two weeks following the transfer in sea water before reaching a new steady state (Bornancin & De Renzis 1972, Thomson & Sargent 1977). The highest activity was seen to correspond to decreasing amounts of plasma electrolytes

(Bornancin & De Renzis 1972). The timing of the increase in the number of chloride cells precedes the increase in the level of ion-pumping activity (Thomson & Sargent 1977) suggesting a maturation phase leading to fully equipped cells at the end of the adaptation period.

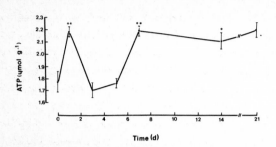

Fig. 4. Gill ATP content in eel transferred from freshwater to sea-water

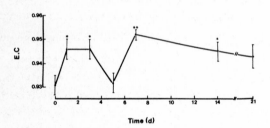

Fig. 5. Gill adenylate energy charge in eel transferred from freshwater to seawater.

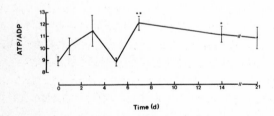

Fig. 6. Gill ATP/ADP ratio in eel transferred from freshwater to sea-water.

Unexpectedly, when the time course of the gill ATP content is observed (fig. 4) at any time its value is lower than the freshwater control value. One day after the seawater transfer and from 6 days to the end of the experiment (3 weeks) the ATP level has significantly higher values than in freshwater control fish. The adenylate pool follows a similar trend during all the experiment. If the high ATP content one day after the transfer can be correlated with a drop in gill Na^+,K^+-ATPase activity and a high plasma Na content observed at the same time (Bornancin & De Renzis 1972), the high steady level of ATP observed after 6 days would be the result of a reduction in energy utilization. The time course changes of the adenylate energy charge (fig. 5) and the ATP/ADP ratio (fig. 6) corroborate also the hypothesis of a high energetic state prevailing in the gill of sea-water transferred eel. The lower value five days after the transfer can be correlated with the development of new chloride cells with increasing amount of membrane ion-pumping activity. Thus the observed adenylate state of the gill in the sea-water eel is not consistant with the expected rise in energy requirement of ion transport. The results from this investigation in eel suggest that, as in trout, sea-water adaptation does not involve at the gill level a further expenditure of energy above the freshwater cost. The difference in behaviour observed in these studies between the gill adenylate metabolism of the two species can be accounted for the difference in their intrinsic adaptative capacities and their ecological origin.

Even if the ionic regulatory mechanism is not the same in fish adapted to marine or freshwater environments (Kirschner 1979), it can be postulated that the difference in energy consumption at the gill level originates mainly from the difference in osmotic work which is directly proportional to ln C1/C2, C1/C2 being the concentration ratio of the ionic species. Inconclusive experiments have been performed on the electrical gradients in fresh- and sea-water fish and too much limitations have been described to draw unequivocal conclusions from these data. Furthermore, the controversy about the active or passive nature of specific ion transport is not yet resolved. It appears therefore that the hyperosmotic regulation may require in fish a higher energy provision than the hypo-osmotic regulation and that metabolic investigations at the gill level are particularly suitable to clear this energetic problem. It seems meaningful that the energy requirements for ion-regulation were calculated to be very similar in a sea-water adapted fish (Potts et al. 1973) and a freshwater fish (Eddy 1975). It would appear that the plausible main mechanism of sea-water adaptation at the gill level is as follows : the increase in plasma ionic concentration due to the drinking reflex initiates regulatory loops involving endocrine relays which induce a new type of cells, the NaCl excreting cells. Their major role in relation with their specialized tubular system is, according to Kirschner (1977), to minimize water loss rather than to extrude Na from blood into an environment only 3 times more concentrated. These two functions water conservation and active salt excretion, likely based on a specific Na,K-ATPase system, cannot be theoretically associated with a higher energy requirement than that required by the freshwater situation where yet unknown structures must transport electrolytes into the blood from an environment 100 to 1000 times less concentrated. It can be postulated that in fish the freshwater state is more costly that it could be estimated from our experiments. This is likely explained by the heterogeneity of the tissue analysed whatever the gill structure involved in ion uptake. The abundance of non-transporting cells (mucus, respiratory, endothelial, nervous) would complicate interpretation of all metabolic data. The opercular or buccal skin preparations of some fish (Karnaky 1980) with a large fraction of chloride cells should be more suitable for such demonstrations. It can be supposed that studies of fish adapted to one third sea-water (isosmotic state) would be of particular interest to assess the energy requirements of ion-regulation in euryhaline fish.

It is hoped that further research developments in the fields of bio-chemistry and physiology would bring new experimental evidences to support our hypothesis concerning gill energetics in euryhaline fish.

II. GILL ENERGETICS IN CRUSTACEA

It is well known that some crustacean species have developed mechanisms of hyperosmotic regulation enabling them to maintain hemolymph concentrations above that of a dilute environment (Kirschner 1979, Lockwood 1977, Gilles & Péqueux 1981). Under these conditions animals reduce the ion and water permeability of the body surface and increase salt uptake at the gill level. Thus, if in sea-water they need a minimal amount of energy since the extracellular fluid and the external fluid are in iso-ionic and isosmotic conditions, the hyperosmotic condition in dilute sea-water requires an expenditure of energy. Among the most euryhaline crabs, *Eriocher sinensis* extends easily its ecological range from sea-water to reasonably soft freshwater. These ecophysiological properties have channelled numerous studies which have established the site of active ion transport (Koch et al. 1954, Gilles & Péqueux 1981), some characteristics of the gill ion pump (Péqueux & Gilles 1977), and the role of amino-acids as intracellular osmotic effectors (Gilles 1979b). To our knowledge, in the field of crustacean gill energetics as a function of salinity only two works are available. Engel et al. (1975) investigated the respiration rates and ATP contents of the blue crab gills at two salinities and Wanson et al. (1983) determined the adenylate pool in the gills of two euryhaline crabs adapted either to full or diluted sea-water, *Carcinus maenas* and *Eriocher sinensis*. The main data from the latter species, a strong regulator, are commented below.

Individual excised gills were processed as fish gills for adenylate analysis but without the need to correct for the hemolymph since this fluid was found to be practically devoid of purine compounds. Despite the scatter of data the posterior set of gills has a two times higher adenylate content than the anterior set in both media   but adaptation does not affect the adenylate content of each set of gills (fig. 7). When the ATP/ADP ratio is considered (fig. 8), it is evident that only in the posterior set of gills, the freshwater adaptation decreases significantly this ratio. In animals acclimated to sea-water, the adenylate energy charge levels off around 0.9 and is slightly higher in posterior gills (table 1). Together with the ATP/ADP ratio, these data are in agreement with an adenylate metabolism running optimally in a steady state where the energy demand remains fairly low. This metabolic state is consistent

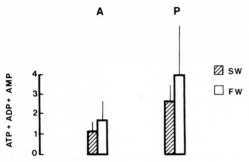

Fig. 7. Gill adenylate concentration (μmol/mg DNA) in Eriocher sinensis adapted to sea-water or freshwater.

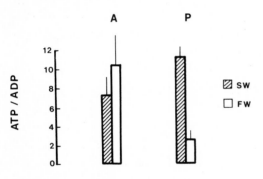

Fig. 8. Gill ATP/ADP ratio in Eriocher sinensis adapted to sea-water or freshwater.

Table 1. Gill adenylate energy charge in Eriocher sinensis adapted to freshwater (FW) or sea-water (SW). Mean values ± SE ; n=9.

Gill	FW	SW	P
Anterior	0.92±0.04	0.87±0.04	0.10
Posterior	0.75±0.07	0.93±0.02	<0.001

with the isosmoticity of the extra-cellular fluid and the surrounding water so that ion-pumping processes are extremely reduced. Adaptation to freshwater results in a significant energy charge drop restricted to the posterior gills. These data suggest that due to an increased energy demand, the ATP production is unable to balance ATP utilization precisely in the epithelial structures known to actively take up Na^+ from the dilute environment (Gilles & Péqueux 1981). The results from these investigations corroborate previous findings on another euryhaline crab (Engel et al. 1975) and demonstrate a close correlation between energetic metabolism and ionic transport. From the positive correlation demonstrated between the rate of oxygen consumption and ATP content in blue crab gills (Engel et al. 1975), it can be postulated that any dilution stress imposing an ionic gradient induces the onset of cellular oxidative metabolism through a tight coupling with ionic transport mechanism (fig. 1). The homogeneous structure of crustacean gills affords the opportunity to analyse more clearly than in fish and probably more precisely than in any other epithelial tissues the coupling of active ion transport to cellular energy production. The various ecophysiological performances found in this invertebrate group offer extensive comparative studies.

Acknowledgement. The author is grateful to Mrs. G. Gutbier for technical assistance.

REFERENCES

Atkinson DE (1977) Cellular energy metabolism and its regulation. Acad Press, New York.

Balaban RS, Mandel LJ, Soltoff S and Storey JM (1980) Coupling of Na-K-ATPase activity to aerobic respiratory rate in isolated cortical tubules from the rabbit kidney. Proc Natl Acad Sci USA 77:447-451.

Blond DM and Whittam R (1980) The regulation of kidney respiration by sodium and potassium ions. Biochem J 92:158-167.

Bornancin M and De Renzis G (1972) Evolution of the branchial sodium outflux and its components, especially the Na-K exchange and the Na-K dependent ATPase activity during adaptation to seawater in *Anguilla anguilla*. Comp Biochem Physiol 43A:577-581.

Colin DA, Kirsch R and Leray C (1979) Haemodynamic effects of adenosine on gills of the trout (*Salmo gairdneri*) J Comp Physiol 130:325-330.

Eddy FB (1975) Effect of calcium on gill potentials and on sodium and chloride fluxes in the goldfish *Carassius auratus*. J Comp Physiol 96:131-142.

Engel DW, Ferguson RL and Eggert LD (1975) Respiration rates and ATP concentrations in the excised gills of the blue crab as a function of salinity. Comp Biochem Physiol 52A:669-673.

Farmer GJ and Beamish WH (1969) Oxygen consumption of *Tilapia nilotica* in relation to swimming speed and salinity. J Fish Res Board Can 26:2807-2821.

Fletcher CR (1975) The energetics of ionic regulation. In : Bolis L, Maddrell HP and Schmidt-Nielsen K (ed.) Comparative Physiology - Functional aspects of structural materials, North-Holland, Amsterdam, p 161.

Giebisch G (1978) Membrane transport in biology, vol. III.Transport across multi-membrane systems. Springer-Verlag, Berlin.

Gilles R (1979a) Mechanisms of osmoregulation in animals. Maintenance of cell volume. John Wiley, New York.

Gilles R (1979b) Intracellular organic osmotic effectors. In : Gilles R (ed.) Mechanisms of osmoregulation in animals.John Wiley,New York, p 111.

Gilles R and Péqueux A (1981) Cell volume regulation in crustaceans : relationship between mechanisms for controlling the osmolality of extracellular and intracellular fluids. J Exp Zool 215:351-362.

Gupta BL, Moreton RB, Oschman JL and Wall BJ (1977) Transport of ions and water in animals. Acad Press, New York.

Holmes WN and Stott GH (1960) Studies of the respiration rates of excretory tissues in the cutthroat trout (*Salmo clarki clarki*). II. Effect of transfer to sea water. Physiol Zool 18:15-20.

Karnaky KJ (1980) Ion-secreting epithelia : chloride cells in the head region of *Fundulus heteroclitus*. Am J Physiol 238:R185-R198.

Kirschner LB (1977) The sodium chloride excreting cells in marine vertebrates. In : Gupta BL, Moreton RB, Oschman JL, Wall BJ (ed.) Transport of ions and water in animals. Acad Press, New-York, p 427.

Kirschner LB (1979) Control mechanisms in crustaceans and fishes. In : Gilles R (ed.), Mechanisms of osmoregulation in animals. John Wiley, New-York, p 157.

Koch JJ, Evans J and Schicks E (1954) The active absorption of ions by the isolated gills of the crab *Eriocher sinensis* (M. EDW.). Konink Vlaamse Akad Wetenschap 16:3-16.

Krogh A (1939) Osmotic regulation in aquatic animals. Cambridge University Press, London.

Leray C, Colin DA and Florentz A (1981) Time course of osmotic adaptation and gill energetics of rainbow trout (*Salmo gairdneri* R.) following abrupt changes in external salinity. J Comp Physiol 144:175-181.

Leray C, Raffin JP and Winninger C (1979) Aspects of purine metabolism in the gill epithelium of rainbow trout, *Salmo gairdneri* Richardson. Comp Biochem Physiol 62B:31-40.

Lockwood APM (1977) Transport and osmoregulation in crustacea. In : Gupta BL, Moreton RB, Oschman JL, Wall BJ (ed.) Transport of ions and water in animals. Acad Press, New York, p 673.

Maloiy GMO (1977) Comparative physiology of osmoregulation in animals. Acad Press, New York.

Mandel LJ and Balaban RS (1981) Stoichiometry and coupling of active transport to oxidative metabolism in epithelial tissues. Am J Physiol 240:F357-F371.

Mayer N and Nibelle J (1970) Kinetics of the mineral balance in the eel *Anguilla anguilla* in relation to external salinity changes and intravascular saline infusions. Comp Biochem Physiol 35:553-566.

Péqueux A and Gilles R (1977) Osmoregulation of the chinese crab *Eriocher sinensis* as related to the activity of the (Na^+-K^+) ATPase. Arch Int Physiol Biochim 85:41-42.

Péquignot J and Serfaty A (1965) Influence de la salinité sur la respiration tissulaire chez les Téléostéens. Experientia 21:227.

Potts WTW (1977) Fish gills. In : Gupta BL, Moreton RB, Oschman JL, Wall BJ (ed.) Transport of ions and water in animals. Acad Press, New York, p 453.

Potts WTW, Fletcher CR and Eddy B (1973) An analysis of the sodium and chloride fluxes in the flounder *Platichthys flesus*. J Comp Physiol

87:21-28.

Raffin JP (1983) Métabolisme énergétique branchial chez les poissons téléostéens : Etude des propriétés de l'AMP déaminase en relation avec quelques facteurs du milieu. Thèse d'Etat, Université de Strasbourg.

Raffin JP and Leray C (1980) Comparative study on AMP deaminase in gill, muscle and blood of fish. Comp Biochem Physiol 67B:533-540.

Rao GMM (1968) Oxygen consumption of rainbow trout (*Salmo gairdneri*) in relation to activity and salinity. Can J Zool 46:781-786.

Silva P, Stoff JS, Solomon RJ, Rosa R, Stevens A and Epstein J (1980) Oxygen cost of chloride transport in perfused rectal gland of *Squalus acanthias*. J Membr Biol 53:215-221.

Stagg RM and Shuttleworth TJ (1982) Na^+,K^+ ATPase, ouabain binding and ouabain-sensitive oxygen consumption in gills from *Platichthys flesus* adapted to seawater and freshwater. J Comp Physiol 147:93-99.

Stuenkel EL and Hillyard SD (1980) Effects of temperature and salinity on gill Na^+,K^+ ATPase activity in the pupfish, *Cyprinodon salinus*. Comp Biochem Physiol 67A:179-182.

Thomson AJ and Sargent JR (1977) Changes in the levels of chloride cells and (Na^+-K^+)-dependent ATPase in the gills of yellow and silver eels adapting to seawater. J Exp Zool 200:33-40.

Ussing HH and Thorn NA (1973) Transport mechanisms in epithelia. Munksgaard, Copenhagen.

Van Rossum GDV (1972) The relation of sodium and potassium ion transport to the respiration and adenine nucleotide content of liver slices treated with inhibitors of respiration. Biochem J 129:427-438.

Wanson S, Péqueux A and Leray C (1983) Effect of salinity changes on adenylate energy charge in gills of two euryhaline crabs. Arch Int Physiol Biochim 91:81-82.

Zerahn K (1956) Oxygen consumption and active sodium transport in the isolated and short-circuited frog skin Acta Physiol Scand 36:300-318.

Regulatory functions of $Na^+ + K^+$-ATPase in marine and estuarine animals

D.W. TOWLE

SUMMARY   Despite many studies which suggest a link between blood Na^+ regulation and the action of $Na^+ + K^+$-ATPase in aquatic animals, the precise role of the ATPase remains unclear. The restricted distribution of $Na^+ + K^+$-ATPase in basolateral membranes of gill cells would promote sodium uptake from the medium, but precludes direct participation of the enzyme in salt secretion. Following alteration of external salt concentration, some euryhaline osmoregulators exhibit adaptive changes in $Na^+ + K^+$-ATPase activity, but others do not. In those animals which do exhibit such changes, short-term modulation of enzymatic activity may be superimposed upon long-term changes in enzyme abundance. In assays of $Na^+ + K^+$-ATPase activity, NH_4^+ substitutes effectively for K^+, but it is unclear whether this biochemical property is related to Na^+/NH_4^+ exchange in intact gill. Multiple functions of the enzyme, possibly including cell volume regulation and generation of Na^+ gradients capable of driving Na^+-coupled transport systems, may obscure its role in transepithelial Na^+ movement. The recent demonstration of ATP-dependent Na^+ transport by basolateral membrane vesicles from crustacean gills may allow an improved estimation of the significance of $Na^+ + K^+$-ATPase in blood Na^+ regulation.

I. INTRODUCTION

The enzymatic equivalent of the sodium pump, sodium-plus-potassium-stimulated adenosine triphosphatase ($Na^+ + K^+$-ATPase), is a nearly ubiquitous, integral component of animal cell membranes. The enzyme serves to maintain low intracellular Na^+ and high intracellular K^+, pumping both ions against concentration gradients. In some highly-specialized epithelial tissues, the prevalence and subcellular distribution of the enzyme have lent support to the conclusion that it is intimately associated with transepithelial ion movements, especially those of Na^+ and Cl^-. In addition, reports of adaptive and timely changes of Na^++K^+-ATPase activity in ion-transporting tissues of osmoregulating aquatic animals have lead to the suggestion that the enzyme plays an important role in whole-body ion regulation by these animals (reviewed by Towle, 1981a). However, questions have arisen in regard to the precise nature of the enzyme's participation in this process, particularly in animals which secrete salt. It is the objective of this review to summarize some controversial aspects of Na^++K^+-ATPase function in marine and estuarine animals and to suggest a new approach to clarification of its role.

II. MOLECULAR NATURE OF Na^++K^+-ATPASE

The enzyme is an integral membrane protein, composed of an alpha subunit (approximate M.W.=100,000) and a beta subunit (approximate M.W.=45,000). Because the alpha subunit contains the catalytic site for ATP hydrolysis at the cytoplasmic side of the plasma membrane, and the binding site for ouabain at the exterior aspect of the membrane, it must be a transmembrane protein. The function of the beta subunit remains unclear. The most commonly-observed subunit composition of Na^++K^+-ATPases from various sources is two alpha: two beta, giving a total molecular weight of the functional transport unit of about 290,000 (reviewed by Cantley, 1981).

Vectorial exchange of Na^+ and K^+ across the plasma membrane is accomplished by attraction of Na^+ ions to high affinity sites near the cytoplasmic surface of the ATPase protein, which itself becomes phosphorylated by ATP. A change in conformation of the protein then exposes the ion-binding sites to the cellular environment. Affinity for Na^+ is reduced and affinity for K^+ is increased, permitting displacement of Na^+ to the outside by extracellular K^+. Upon binding of K^+, the protein is dephosphorylated, returning to its original conformation and bringing K^+ into the cellular interior. In many cells, hydrolysis of one ATP molecule energizes the exchange of three

Na⁺ ions for two K⁺ ions, resulting in low Na⁺ concentration, high K⁺ concentration, and electronegativity within cells (Cantley, 1981).

III. LOCALIZATION OF NA⁺+K⁺-ATPASE AND IMPLICATIONS FOR TRANSPORT

In the plasma membrane of most salt-transporting epithelial cells, including those of teleost and crustacean gill, Na⁺+K⁺-ATPase molecules are not distributed randomly but rather exhibit a distinct polarity, being restricted to basolateral regions of the membrane (Hootman and Ernst, this volume; Towle et al., 1983). The relative lack of Na⁺+K⁺-ATPase molecules in the apical membrane, coupled with a prevalence of the enzyme in greater quantities in the basolateral membrane than required to maintain cellular ionic ratios, permits the generation of transepithelial as well as transmembrane ionic gradients. Basolateral Na⁺+K⁺-ATPase would move Na⁺ from the cytosol toward the extracellular fluid in exchange for a counterion, usually assumed to be K⁺. Because substantial ATP-dependent Na⁺ expulsion would not occur at the apical membrane, the directional transfer of

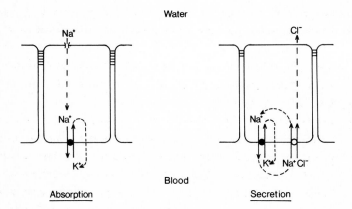

Figure 1. Model of NaCl absorption and secretion by branchial epithelia of aquatic animals. According to this model, basolateral Na⁺+K⁺-ATPase exports Na⁺ across the basolateral membrane in exchange for K⁺, and Na⁺-coupled Cl⁻ transport across the basolateral membrane permits transepithelial NaCl secretion.

Na$^+$ across the epithelium is made likely, resulting in net absorption of Na$^+$ from the apical medium (Figure 1).

The participation of Na$^+$+K$^+$-ATPase in ion secretion across gills or salt glands of hyporegulators must be indirect, because of its directional and locational polarity. That is, the enzyme is capable only of moving Na$^+$ from cytosol to cell exterior, and must do so only across the basolateral membrane of these epithelia. Currently under active investigation is the possibility that the Na$^+$ gradient generated by Na$^+$+K$^+$-ATPase may energize a Na$^+$-coupled Cl$^-$ uptake from the blood across the basolateral membrane (Zadunaisky, this volume). The intracellular Cl$^-$ would then be expelled across the apical membrane according to its electrochemical gradient, drawing Na$^+$ with it and resulting in net secretion of NaCl (Figure 1). The major difference between uptake and secretion would thus be the unique presence of Na$^+$-coupled Cl$^-$ transport in basolateral membranes of epithelia capable of salt secretion.

IV. RESPONSE OF NA$^+$+K$^+$-ATPASE TO CHANGING SALINITY

Many epithelia which demonstrate high rates of NaCl transport possess high Na$^+$+K$^+$-ATPase activity (Bonting, 1970). For example, posterior gills of hyperregulating crabs are strong transporters of Na$^+$ and also demonstrate potent enzyme activity (Pequeux and Gilles, 1981). The distribution of Na$^+$+K$^+$-ATPase activity within the crustacean gill epithelium corresponds well with the distribution of mitochondria-rich "ion-transporting" cells (Copeland and Fitzjarrell, 1968; Neufeld et al., 1980).

When hyperregulating aquatic animals are subjected to reduced salinity, Na$^+$+K$^+$-ATPase activity in gills tends to increase in an adaptive fashion. That is, in many species, as the blood-medium difference in salt concentration becomes larger, enzyme activity in the transporting tissue increases. Such a response has been demonstrated recently in posterior gills of the shore crab Carcinus maenas (Siebers et al., 1982) (Figure 2), confirming previous observations of salinity/enzyme activity relationships in many hyperregulating vertebrates and invertebrates (reviewed by Towle, 1981a).

Marine teleosts, elasmobranchs, and a few marine crustaceans are among the aquatic animals which demonstrate hyporegulation of blood

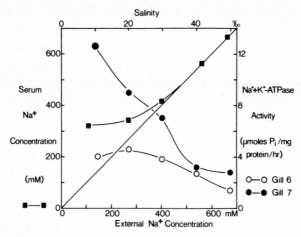

Figure 2. Blood Na$^+$ regulation and Na$^+$+K$^+$-ATPase activity in gills of green crab, <u>Carcinus maenas</u>, in relation to environmental salinity. Note the difference between enzymatic activities of posterior gills (exemplified by gill 7) and anterior gills (exemplified by gill 6). Redrawn from Siebers et al., 1982.

ions. When the salinity is increased from a value isoionic to the blood of euryhaline species within this group of animals, Na$^+$+K$^+$-ATPase activity of salt-secreting tissues tends to increase, apparently supporting enhanced NaCl secretion. For example, seawater-acclimated killifish (<u>Fundulus heteroclitus</u>) exhibit greater gill Na$^+$+K$^+$-ATPase activity than killifish acclimated either to freshwater or to an intermediate salinity near their isoionic value (Epstein et al., 1967; Towle et al., 1977). Killifish acclimated to an isoionic salinity demonstrated the lowest enzyme activity, lending support to the idea that the stress of either hyporegulation or hyperregulation is met, at least in part, by an enhancement of basolateral Na$^+$+K$^+$-ATPase activity (Towle et al., 1977). Similar patterns of adaptation have been observed in a number of teleosts (Kirschner, 1980; Towle, 1981a).

In some teleost species, however, Na$^+$+K$^+$-ATPase activity may not demonstrate easily-interpretable responses to salinity change. In the flounder <u>Platichthys flesus</u>, for example, enzyme activity calculated on the basis of wet gill weight showed no difference between freshwater- and seawater-acclimated animals (Stagg and Shuttleworth, 1982). Calculation on the basis of milligrams protein, which may relate more closely to the function of transport enzymes, showed that freshwater-acclimated flounder exhibited higher ATPase activity. Data for animals acclimated to an isoionic salinity were not presented. In any case, the freshwater/seawater dichotomy in levels of ATPase activity is not universal, the differences between species perhaps reflecting evolutionary origins in marine or freshwater environments (Stagg and Shuttleworth, 1982).

In those animals which demonstrate adaptive changes in gill $Na^+ + K^+$-ATPase activity, short-term modulation of enzyme activity may be superimposed on long-term effects of cell restructuring and membrane proliferation. In the blue crab <u>Callinectes sapidus</u>, for example, gill ATPase activity increases abruptly within a few hours after salinity reduction, but then requires nearly two weeks to reach a steady-state level (Towle et al., 1976; Neufeld et al., 1980). The early modulation of enzyme activity may be the result of direct activation of existing but latent enzyme molecules, or may result from rapid reduction in the rate of ATPase degradation, permitting ATPase molecules to accumulate in the plasma membrane. The long-term increase in $Na^+ + K^+$-ATPase activity in crab gill is undoubtedly the result of proliferation of ion-transporting cells within the gill epithelium, with its accompanying production of extensive basolateral membrane (Copeland and Fitzjarrell, 1968). Similar kinds of short- and long-term changes in ATPase activity in relation to environmental salinity have been observed in <u>F. heteroclitus</u> and in the fiddler crab <u>Uca</u> (Karnaky et al., 1976; Towle et al., 1977; Holliday, personal communication).

V. AMMONIUM ION AS COUNTERION

External application of amiloride inhibits both Na^+ uptake and NH_4^+ secretion by the gills of some euryhaline animals (Kormanik and Cameron, 1981; Pressley et al., 1981), supporting the idea that these two processes are coupled at the apical membrane. However, NH_4^+ substitutes quite effectively for K^+ in <u>in vitro</u> assays of crustacean and teleost $Na^+ + K^+$-ATPase (Skou, 1960; Towle and Taylor, 1976; Bell et al., 1977; Mallery, 1983) (Figure 3), suggesting that the basolateral $Na^+ + K^+$-ATPase mediates gill Na^+/NH_4^+ exchange (Mangum and Towle, 1977). The evidence remains unclear. For example, when the isolated head of the marine teleost <u>Opsanus beta</u> was perfused with ouabain,

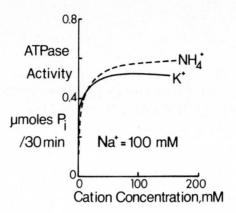

Figure 3.   Effect of K^+ and NH_4^+ on activity of Na^++K^+-ATPase in nerve membranes from the green crab, <u>Carcinus maenas</u>.   Redrawn from Skou, 1960.

ammonia efflux decreased by 50%, apparently the result of inhibition of basolateral Na^++K^+-ATPase (Claiborne et al., 1982). However, perfusion pressure was also altered by ouabain treatment, resulting in a problematic interpretation of the ammonia efflux data. Hemodynamic alterations following ouabain infusion have in fact plagued a number of studies designed to describe the function of Na^++K^+-ATPase in fish gill ion transport. The simpler circulation patterns in crustacean gills would appear to provide an alternative in studies using perfused ouabain, but, unfortunately, the Na^++K^+-ATPase of crustacean tissues is notably insensitive to ouabain, some 100-fold less sensitive than other arthropod and most vertebrate enzymes (Towle et al., 1982).

VI.   OTHER FUNCTIONS OF NA^++K^+-ATPASE

Na^++K^+-ATPase in gills of euryhaline animals may serve functions in addition to maintenance of intracellular ionic ratios and transepithelial Na^+ movements. For example, cellular volume regulation may depend on adjustments in total intracellular ions as well as free amino acids. Na^++K^+-ATPase would be expected to have an important role in such ion-mediated volume regulation by tissues exposed to a changing salinity (Evans, 1979). However, in a study of ATPase activities in the three major tissues within the branchial chamber of the coconut crab _Birgus latro_, only the gills exhibited substantial $Na^+ + K^+$-ATPase activity. Pericardial sac and chamber lining ("lung"), although exposed to the same osmotic stresses as the gill, showed very low activities (Towle, 1981b). The ATPase may indeed assist in volume regulation, but these data suggest that volume regulation cannot be its sole function.

More likely is a secondary function of $Na^+ + K^+$-ATPase in generating transmembrane Na^+ gradients, the potential energy of which drives the cellular uptake of other ions, particularly Cl^-, Ca^{+2}, and Mg^{+2}, and of organic molecules (Holliday and Miller, 1980; Roer, 1980; Zadunaisky, this volume). In tissues exhibiting these Na^+-coupled processes, Na^+ transport rates across the epithelium would not necessarily be correlated with $Na^+ + K^+$-ATPase activity. Rather, the total of the Na^+ transport rate itself plus all Na^+-coupled transport rates, assuming known stoichiometries, might correlate more closely with the measured $Na^+ + K^+$-ATPase activity.

VII. A NEW APPROACH TO STUDYING $Na^+ + K^+$-ATPASE FUNCTION IN EPITHELIA OF EURYHALINE ANIMALS

Measured $Na^+ + K^+$-ATPase activities in homogenized tissues probably do not represent actual rates of transepithelial Na^+ movement, although correlations have been noted (Forrest et al., 1973). Even rates of basolateral Na^+ transport alone may not be reflected in the measured enzyme activity. An independent means of determining the transport function of $Na^+ + K^+$-ATPase is necessary. Such a determination may be permitted by the use of basolateral membrane vesicles isolated from transporting tissues. In such preparations, the functional polarity of epithelial cells is no longer a complicating factor, since basolateral processes can be studied apart from apical processes. Composition of the medium on each side of the vesicular membrane can be controlled easily, and the vesicles themselves can be manipulated with conventional biochemical methods. Separation of intravesicular from free molecules is easily effected by column separation or ultrafiltration (Murer and Kinne, 1980).

We have applied such an approach to a study of Na^+ transport by

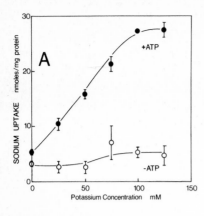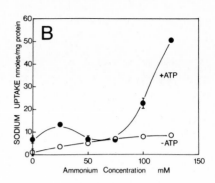

Figure 4. Effect of K^+ (A) and NH_4^+ (B) on Na^+ uptake by basolateral vesicles from posterior gills of the blue crab, <u>Callinectes</u> <u>sapidus</u>. Intravesicular medium contained 10 mM Tris-HEPES (pH 7.4), KCl or NH_4Cl, plus isosmotic sucrose. Extravesicular medium contained 10 mM Tris-HEPES (pH 7.4), 4 mM $^{22}NaCl$, 2 mM $MgCl_2$, 300 mM sucrose, and 4 mM Tris-ATP as indicated. After a 2-minute incubation, vesicles were rapidly filtered, washed with 10 mM Tris-HEPES (pH 7.4) plus 150 mM NaCl, and counted in Aquasol (Fuhrman et al., 1983).

basolateral membrane vesicles from the posterior gills of the blue crab <u>C. sapidus</u>. Regions of gill lamellae highly enriched in ion-transporting cells are used as the starting material, and inside-out vesicles are isolated by a rapid two-step centrifugation technique (Towle et al., 1983; Boumendil-Podevin and Podevin, 1983). Sodium uptake by the vesicles is dependent on the presence of ATP in the extravesicular medium, and on the presence of a counterion in the intravesicular medium. Interestingly, at low concentrations, NH_4^+ substitutes effectively for K^+ as a counterion for Na^+ uptake (Fuhrman et al., 1983) (Figure 4). The complex response of ATP-dependent Na^+ uptake to intravesicular NH_4^+ is quite reproducible (Figure 4B).

Our early results with this experimental system support a model of gill Na^+ regulation in which Na^+/NH_4^+ exchange is a basolateral process, catalyzed by Na^++K^+-ATPase (Figure 5). In the absorptive mode, coupling between Na^+ intake and NH_4^+ output at the apical membrane would be indirect, by way of the basolateral pump. In addition, the stoichiometry of Na^+/NH_4^+ exchange could vary, depending on the relative availability of K^+ and NH_4^+ to the basolateral trans-

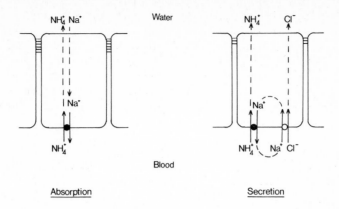

Figure 5. Model of NaCl absorption and secretion by branchial epithelia of aquatic animals, with NH_4^+ serving as an alternative counterion for basolateral Na^+ transport by $Na^+ + K^+$-ATPase.

porting system. In the secretory mode, any linkage between the export of NH_4^+ and that of Cl^- would be even more indirect, by way of the basolateral pump and the Na^+-coupled Cl^- transport system. Demonstration of NH_4^+ transport and Na^+/Cl^- cotransport in isolated basolateral membranes from secretory epithelia should allow an assessment of this model. In addition, isolation of apical membranes in vesicular form will permit an analysis of possible exchange processes going on in that membrane domain as well. It is still not clear whether $Na^+ + K^+$-ATPase serves as the controlling factor in transepithelial NaCl movements in aquatic animals, or more simply provides a limiting ceiling on maximum transport rates.

ACKNOWLEDGEMENTS

Many students have contributed to the work summarized here. The most recent of these are Linda Fuhrman, Brent Stansbury, and Todd Kays. The work is supported by Jeffress Memorial Trust, National Science Foundation (PRM-8213306), and University of Richmond Faculty Research Program.

REFERENCES

Bell MV, Tondeur F and Sargent JR (1977) The activation of sodium-plus-potassium ion-dependent adenosine triphosphatase from marine teleost gills by univalent cations. Biochem. J. 163: 185-187.

Bonting SL (1970) Sodium-potassium activated adenosinetriphosphatase and cation transport. In: Bittar EE (ed.) Membranes and ion transport vol. 1. John Wiley and Sons, London, pp. 257-363.

Boumendil-Podevin EF and Podevin RA (1983) Effects of ATP on Na^+ transport and membrane potential in inside-out renal basolateral vesicles. Biochim. Biophys. Acta 728: 39-49.

Cantley LC (1981) Structure and mechanism of the (Na,K)-ATPase. In: Sanadi DR (ed.) Current topics in bioenergetics vol. 11. Academic Press, New York, pp. 201-237.

Claiborne JB, Evans DH and Goldstein L (1982) Fish branchial Na^+/NH_4^+ exchange is via basolateral Na^+-K^+-activated ATPase. J. Exp. Biol. 96: 431-434.

Copeland DE and Fitzjarrell AT (1968) The salt absorbing cells in the gills of the blue crab (Callinectes sapidus Rathbun) with notes on modified mitochondria. Z. Zellforsch. 92: 1-22.

Epstein FH, Katz AI and Pickford GE (1967) Sodium- and potassium-activated adenosine triphosphatase of gills: role in adaptation of teleosts to salt water. Science 156: 1245-1247.

Evans DH (1979) Fish. In: Maloiy GMO (ed.) Comparative physiology of osmoregulation in animals vol. 1. Academic Press, New York, pp. 305-390.

Forrest JN Jr, Cohen AD, Schon DA and Epstein FH (1973) Na transport and Na-K-ATPase in gills during adaptation to seawater: effects of cortisol. Am. J. Physiol. 224: 709-713.

Fuhrman LA, Stansbury BR and Towle DW (1983) ATP-Dependent Na^+ transport by basolateral membrane vesicles from crab gill. Am. Zool. 23: (Abst).

Holliday CW and Miller DS (1980) PAH transport in rock crab (Cancer irroratus) urinary bladder. Am. J. Physiol. 238: R311-R317.

Karnaky KJ Jr, Kinter LB, Kinter WB and Stirling CE (1976) Teleost chloride cell II. Autoradiographic localization of gill Na,K-ATPase in killifish Fundulus heteroclitus adapted to low and high salinity environments. J. Cell Biol. 70: 157-177.

Kirschner LB (1980) Comparison of vertebrate salt-excreting organs. Am. J. Physiol. 238: R219-R223.

Kormanik GA and Cameron JN (1981) Ammonia excretion in animals that breathe water: a review. Mar. Biol. Lett. 2: 11:23.

Mallery CH (1983) A carrier enzyme basis for ammonium excretion in teleost gill. NH_4^+-stimulated Na-dependent ATPase activity in Opsanus beta. Comp. Biochem. Physiol. 74A: 889-897.

Mangum C and Towle D (1977) Physiological adaptation to unstable environments. Am. Sci. 65: 67-75.

Murer H and Kinne R (1980) The use of isolated membrane vesicles to study epithelial transport processes. J. Membrane Biol. 55: 81-95.

Neufeld GJ, Holliday CW and Pritchard JB (1980) Salinity adaption of gill Na,K-ATPase in the blue crab, Callinectes sapidus. J. Exp. Zool. 211: 215-224.

Pequeux A and Gilles R (1981) Na^+ fluxes across isolated perfused gills of the Chinese crab Eriocheir sinensis. J. Exp. Biol. 92: 173-186.

Pressley TA, Graves JS and Krall AR (1981) Amiloride-sensitive ammonium and sodium ion transport in the blue crab. Am. J. Physiol. 241: R370-R378.

Roer RD (1980) Mechanisms of resorption and deposition of calcium in the carapace of the crab Carcinus maenas. J. Exp. Biol. 88: 205-218.

Siebers D, Leweck K, Markus H and Winkler A (1982) Sodium regulation in the shore crab Carcinus maenas as related to ambient salinity. Mar. Biol. 69: 37-43.

Skou JC (1960) Further investigations on a $Mg^{++}+Na^+$-activated adenosine triphosphatase, possibly related to the active, linked transport of Na^+ and K^+ across the nerve membrane. Biochim. Biophys. Acta 42: 6-23.

Stagg RM and Shuttleworth TJ (1982) Na^+,K^+ ATPase, ouabain binding and ouabain-sensitive oxygen consumption in gills from Platichthys flesus adapted to seawater and freshwater. J. Comp. Physiol. 147: 93-99.

Towle DW (1981a) Role of Na^++K^+-ATPase in ionic regulation by marine and estuarine animals. Minireview. Mar. Biol. Lett. 2: 107-122.

Towle DW (1981b) Transport-related ATPases as probes of tissue function in three terrestrial crabs of Palau. J. Exp. Zool. 218: 89-95.

Towle DW, Gilman ME and Hempel JD (1977) Rapid modulation of gill Na^++K^+-dependent ATPase activity during acclimation of the killifish Fundulus heteroclitus to salinity change. J. Exp. Zool. 202: 179-186.

Towle DW, Kays WT and Cioffi M (1983) Localization of $Na^+ + K^+$-ATPase in basolateral membranes of crab gill ion-transporting cells. Am. Zool. 23: (Abst).

Towle DW, Mangum CP, Johnson BA and Mauro NA (1982) The role of the coxal gland in ionic, osmotic, and pH regulation in the horseshoe crab <u>Limulus polyphemus</u>. In: Bonaventura J, Bonaventura C and Tesh S (eds.) Physiology and biology of horseshoe crabs: studies on normal and environmentally-stressed animals. Alan R. Liss Inc., New York, pp. 147-172.

Towle DW, Palmer GE and Harris JL III (1976) Role of gill $Na^+ + K^+$-dependent ATPase in acclimation of blue crabs (<u>Callinectes</u> <u>sapidus</u>) to low salinity. J. Exp. Zool. 196: 315-322.

Towle DW and Taylor DD (1976) Effect of NH_4^+ and K^+ on Na^+-transport ATPase activity of blue crab gill. Am. Zool. 16: 224 (Abst).

Ultrastructural localization of $Na^+ + K^+$-ATPase in specialized membranes of salt transporting cells in marine vertebrates

S.R.HOOTMAN and S.A.ERNST

All marine vertebrates regulate the salt concentration of their body fluids to levels far below those found in the marine environment. In addition to renal mechanisms, osmotic homeostasis is maintained by the active excretion of NaCl from specialized organs, the location and structure of which varies considerably among the classes of vertebrates. Thus in marine teleosts, the gill is the primary site of extra-renal salt secretion, while in elasmobranch fish the rectal gland fulfills this role. In marine birds, paired glands located just above the orbits of the eyes are responsible for salt secretion. Considering the phylogenetic diversity and varied anatomical location of these organs, one might expect that the secretory cells that comprise each of these glands would exhibit a comparable diversity of structure. This, however, is not the case. Rather, the commonality of function shared by these cells is mirrored in the underlying similarity of their cytoarchitectural design. The most salient features shared by these epithelial cells are an extensive elaboration of basolateral plasma membranes and an abundant supply of mitochondria, the mitochondria generally in close proximity to the amplified basal and lateral cell surfaces. Other structural features, such as specific modifications of the occluding junctions that unite adjacent secretory cells, also exhibit marked similarities in these salt secretory tissues.

Appreciation of these morphological similarities inevitably suggested that they would be mirrored by similarities in transport components resolved at the biochemical level. The results of numerous investigations now have implicated one enzyme as a key component in the biochemical machinery that subserves electrolyte transport in salt secreting tissues. This enzyme is the sodium, potassium-activated adenosine triphosphatase (Na,K-ATPase). Levels of Na,K-ATPase activity in epithelia of hypertonic salt secreting glands are much higher than those measured in most other epithelia and in non-transporting tissues. In addition, in some organs such as the avian salt gland and gills of teleost fish, an increase in functional demand (i.e., salt stress) results in a significant rise in tissue levels of Na,K-ATPase, a rise that parallels specific modifications in the structure of the secretory cells.

These correlations led us several years ago to initiate a course of studies designed to localize sites of Na,K-ATPase activity in selected salt secreting tissues and thereby to approach an understanding of its involvement in the secretory process. In this article, we review some of this work, focusing on the specific morphological adaptations common to salt glands in general and on the localization of Na,K-ATPase within these glands. We will limit our observations to three organs; the avian salt gland, the elasmobranch rectal gland, and the teleost gill/operculum. To conclude, we will integrate these observations into a model of salt gland function that reflects our present state of knowledge and provides a logistical foundation for further studies.

I. Techniques for Ultrastructural localization of Na, K-ATPase

There are at present several techniques available that provide valid localization of Na,K-ATPase. These include autoradiographic localization of ^{3}H-ouabain binding sites (Stirling, 1976; Ernst and Mills, 1980), immunocytochemical localization using antibodies to the holoenzyme or catalytic subunit and ferritin- or peroxidase-labelled secondary antibodies (Kyte, 1976a,b; Schwartz, et al., 1981), and cytochemical localization using para-nitrophenylphosphate (ρ-NPP) as substrate and either lead chelated with citrate (Mayahara, et al., 1980) or strontium (Ernst, 1972a,b) as capture ion. Localization of ^{3}H-ouabain binding sites is a highly specific and quantitative technique, but is restricted to the light microscopic level and is of utility only in those tissues where the enzyme-ouabain complex is of high affinity ($K_d < 1$ μM). Conversely, immunocytochemical procedures provide localizations with very high sensitivity but are often limited by poor penetration of immunoreactants to membrane sites and by the necessity of purifying the Na,K-ATPase and raising antibodies to it. In addition, since immunocytochemical localization is based on antigenicity alone, it does not discriminate between active and inactive enzyme complexes. Consequently, the cytochemical approaches have been used most widely to localize sites of Na,K-ATPase activity at the ultrastructural level. Both procedures give complementary results when applied to the same tissues (Ernst, 1975; Mayahara and Ogawa, 1980). The

Mayahara technique is somewhat simpler to carry out, while the Ernst technique on which it is based appears to be more sensitive in certain instances (Ernst and Schreiber, 1981). We recently reviewed the advantages and limitations of these cytochemical techniques and of the autoradiographic and immunocytochemical approaches as well (Ernst and Hootman, 1981). Data obtained from several tissues in which all of these methods have been employed are complementary, thereby establishing cross-validation of these procedures. In the following sections, we will present results obtained primarily using the Sr-NPPase technique which was developed in our laboratory (Ernst, 1972a,b).

The Sr-NPPase procedure was developed to circumvent the many problems that frustrated attempts to localize Na,K-ATPase activity in tissues using the classical Wachstein-Meisel ATPase procedure. The NPPase technique makes use of the fact that Na,K-ATPase will hydrolyze ρ-NPP in a reaction that is equivalent to the dephosphorylation step of the complete reaction sequence of the enzyme (Dahl and Hokin, 1974). This partial reaction is both K^+-dependent and ouabain-sensitive, thus providing opportunities for the inclusion of unambiguous controls in each cytochemical experiment. In addition, the reaction provides two products, the nitrophenolate ion, which can be quantitated spectrophotometrically, and inorganic phosphate, which is precipitated at the site of enzymatic release by strontium.

In practice, localizations are obtained by incubating thin sections (< 100 μM) of tissues fixed briefly with formaldehyde or formaldehyde-glutaraldehyde in a buffered medium containing ρ-NPP, $SrCl_2$, $MgCl_2$, and KCl and then rinsing the tissues in a solution of $Pb(NO_3)_2$ to convert precipitated Sr phosphate to the more electron dense Pb salts. The tissue sections are then postfixed in OsO_4, embedded in epoxy resins, and examined by conventional transmission electron microscopic techniques. Membrane associated precipitates that are seen in thin sections, and are abolished when K^+ is omitted from or ouabain added to the reaction medium are considered to represent sites of Na,K-ATPase activity.

II. Ultrastructure of salt secreting tissues and localization of $Na^+ + K^+$-ATPase

A. Avian salt gland

Salt glands are present in most birds that ingest significant quantities of NaCl in their diet or drinking water. These paired glands secrete a highly saline effluent via excretory ducts that open at the external nares. The glands consist of numerous lobules of secretory tissue, each of which is in turn composed of radially arranged branched tubules that

open into centrolobular collecting ducts which anastomose to form the
common excretory duct. Each tubule is composed of a single layer of epithelial cells. When the rate of glandular secretion is low, as when birds
are chronically drinking fresh water, the secretory cells display only a
moderate degree of development. At the ultrastructural level, the plasma
membrane which rests on the encompassing basal lamina exhibits only a
minor amount of folding (Ernst and Ellis, 1969; Martin and Philpott,
1973; Hootman and Ernst, 1980). In contrast, the lateral cell surfaces
display numerous membrane folds or plicae which extend into the prominent
intercellular spaces between adjacent secretory cells. In birds adapted
to conditions of high salt intake (Fig. 1), the basal cell surface is

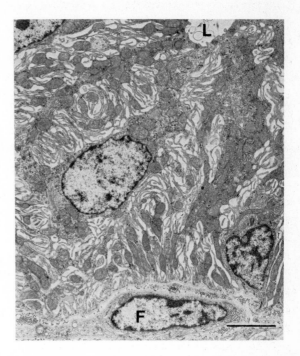

Figure 1. Salt gland of the
domestic duck Anas platyrhynchos given 1% NaCl to drink
for two weeks. Epithelial
cells are characterized by
extreme folding of basolateral plasma membranes. Mitochondria are intimately
associated with these folds.
Apical surfaces of adjacent
secretory cells which delimit
the tubular lumen (L) display
a few stubby microvilli.
Fibroblast (F). Scale bar is
 3 μm.

vastly expanded into a labyrinth of drapery-like folds of plasma membrane
that extend from the cell cytocenter to the basal lamina (Ernst and Ellis,
1969). The folds characteristically enclose numerous mitochondria, an
adaptation that presumably increases the efficiency of energy transfer
from these organelles to sites of ion transport activity.

A second modification of cell structure induced by chronic salt stress
is seen when the occluding junctions that unite adjacent secretory cells
are examined. In micrographs of freeze-fractured tissue replicas (Ellis,
et al., 1977; Riddle and Ernst, 1979), the occluding junctions in birds
adapted to a low salt stress situation typically consist of one or two
roughly parallel pairs of closely juxtaposed strands (on the P fracture

face) or grooves (on the E face). When present as two doublets, each pair is separated by a distance of about 20 nm, the complete junction thus consisting of four strands with a total depth of 60-80 nm. In birds whose salt intake is high, only one pair of closely apposed parallel strands or grooves is seen (Fig. 2). This figure also illustrates a

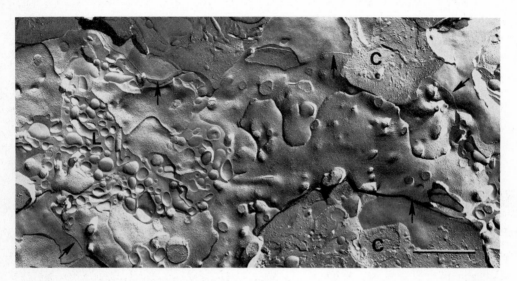

Figure 2. Freeze-fractured replica of the luminal surface of epithelial secretory cells from salt gland of a domestic duck adapted to 1% NaCl. Portions of the tubular lumen (L) and cross-fractured cytoplasm (C) of cells are visible. Cross-fractured bases of microvilli characterize the luminal surface. Occluding junctions, consisting in most instances of two parallel strands, are seen in several places (arrows). Note also the interlocking nature of lateral margins of adjacent secretory cells, as seen at center. Scale bar is 1 μm.

second notable feature of the junction in the salt gland. In en face views of the fractured apical surface, it is apparent that the lateral margins of secretory cells are highly folded, a structural modification of the epithelium that increases the linear extent of occluding junction several-fold over that which a non-folded lateral cell margin would afford. Morphometric analysis (unpublished data) shows that approximately 60 meters of junctional length is present per cm^2 of apical surface.

Early biochemical studies on the salt glands of the herring gull and other marine birds (Bonting, et al., 1964) showed them to be rich in Na,K-ATPase activity, surpassing almost all other tissues that had to that time been investigated. Not long thereafter, we demonstrated that

Na,K-ATPase activity in salt glands of domestic ducks was strictly correlated to the degree of salt stress experienced by the animals (Ernst,
et al., 1967). When ducklings given fresh water to drink were switched
to 1% NaCl, Na,K-ATPase activity in the salt glands was elevated by 24
hours, and had increased to a level approximately five-fold higher than
control values by seven days. More recent studies have clearly shown that
this increase in ATPase activity is due to de novo synthesis of the enzyme (Stewart, et al., 1976). Correlation of increased synthesis of Na,
K-ATPase molecules with the morphologically demonstrable increase in
folding of basolateral plasma membranes in the secretory cells suggested
that the new enzyme units were being inserted into this expanding cell
surface. Unequivocal demonstration that this was, in fact, the case was
obtained by application of the Sr-NPPase localization procedure to tissue
sections of salt-stressed duck salt gland (Ernst, 1972a,b). Results from
such an experiment are shown in Fig. 3. Electron dense precipitates indi-

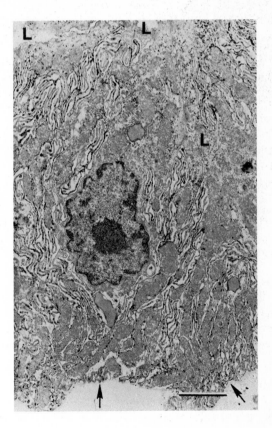

Figure 3. Sr-NPPase cytochemical localization of Na,K-ATPase
in salt gland of a duck adapted
to 1% NaCl. Precipitates indicative of enzyme activity line
the cytoplasmic surfaces of
folds of basolateral plasma membrane. Apical plasma membranes
bordering the tubular lumen (L)
are devoid of reaction products.
Basal lamina (arrows). Scale
bar is 2 μm.

cative of sites of Na,K-ATPase activity are localized to the cytoplasmic
surfaces of the folds of basolateral plasma membrane that are developed

by the secretory cell in response to salt stress. The distribution of
precipitates is uniform along the lengths of these folds. Precipitates
are sparse or absent on the apical plasma membrane that borders the luminal compartment. Although not shown here, micrographs of tissue sections
incubated in reaction media either lacking K^+ or Mg^{++} or containing 10
mM ouabain show a dramatic reduction in precipitate deposition (Ernst,
1972b). These results clearly establish that the reaction products seen
in Fig. 3 represent sites of Na,K-ATPase activity. This localization
also has been corroborated in this laboratory by autoradiographic studies
using ^{3}H-ouabain (Ernst and Mills, 1977).

B. Elasmobranch rectal gland

Elasmobranch fish (sharks, skates, and rays) possess a unique digiform
salt gland that is suspended in the peritoneal mesentery and opens via an
excretory duct into the cloaca. The ability of the gland to concentrate
and secrete a hypersaline effluent was first documented by Burger and
Hess (1960). The first detailed anatomical and ultrastructural studies of
the gland appeared shortly thereafter (Bulger, 1963). The parenchyma of
the gland consists of epithelial tubules radially arranged about a central collecting duct, an organization reminiscent of that seen in individual lobules of the avian salt gland. Each tubule consists of a single
layer of cuboidal-to-columnar cells (Fig. 4). These cells are character-

Figure 4. Rectal gland of a dogfish shark. Tissue was postfixed
with osmium-ferrocyanide to en-
hance membrane contrast. Secretory cells contain many mitochondria and an extensively folded
basolateral plasma membrane. The
cell surfaces bordering the lum-
en (L) display short microvilli.
Adjacent cells overlap and in-
terdigitate with each other,
resulting in large numbers of
occluding junction profiles (arrows) along short lengths of
luminal epithelial surface.
Scale bar is 4 µm.

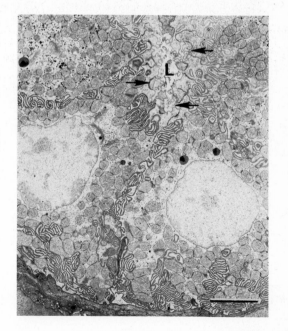

ized by extensive plasma membrane folding along their basal and lateral margins and by an abundance of mitochondria. A striking feature of the luminal epithelial surface, as seen in transmission electron micrographs, is the high incidence of cross-sectioned profiles of occluding junction. When quantitated morphometrically (Ernst, et al., 1981), it is clear that the linear extent of junction in this epithelium (60-80 m/cm^2) is as much as an order of magnitude greater than in many other epithelia, particularly when compared to those that express low transepithelial ionic conductivity. Three-dimensional reconstructions of tubule cells from light micrographs (Bulger, 1963) or images of freeze-fractured replicas (Ernst, et al., 1981) demonstrate that, as in the avian salt gland, this situation results from the presence of interlocking extensions of the lateral borders of adjacent secretory cells.

The structure of occluding junctions in the secretory parenchyma of the rectal gland also is similar to that seen in the avian salt gland. In both the stingray Dasyatis sabina and dogfish Squalus acanthias, junctions consist of closely apposed parallel strands (Fig. 5). Strand number

Figure 5. Freeze-fractured replica of the luminal epithelial surface of a rectal gland tubule from the dogfish. Occluding junctions represented by closely juxtaposed strands on fractured membrane surfaces are visible at several points (arrows). These separate apical (Ap) from lateral (La) cell surfaces. Junctional strand number in this micrograph ranges from two to four. Tubular lumen (L). Scale bar is 1 μm.

is more variable, however. In the stingray, from one to three strands were observed, while in the dogfish as few as two or as many as five strands were seen (Ernst, et al., 1981).

In common with the avian salt gland, the rectal gland is a rich source of Na,K-ATPase (Hokin, et al., 1973). Complementary results have been obtained for localization of the enzyme by both autoradiographic (Eveloff, et al., 1979) and cytochemical (Goertemiller and Ellis, 1976) means. Cytochemical localization utilizing the Sr-NPPase technique (Fig. 6) showed abundant deposition of precipitates on the folded lateral and basal plasma membranes. Reaction product was restricted to the cytoplasmic plasma membrane surface and could be abolished by inclusion of 1 mM ouabain in the reaction medium or omission of K^+. The luminal surface of the epithelium was essentially devoid of reaction products.

Figure 6. Ultrastructural localization of Na,K-ATPase activity in dogfish rectal gland by the Sr-NPPase cytochemical procedure. Precipitates line the inner surfaces of lateral and basal plasma membranes. Reaction product deposition is moderately heavy and uniformly distributed along the length of the cells, but is sparse at the apical cell surface bordering the tubule lumen (L). Micrograph courtesy of Dr. R. A. Ellis, by permission of Springer-Verlag. Scale bar is 3 μm.

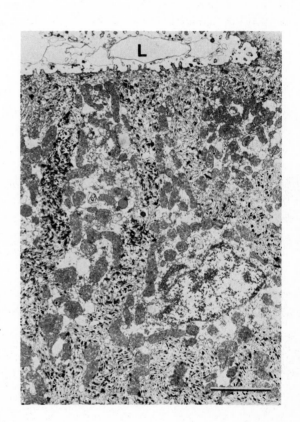

C. Teleost gill/operculum

Unlike the avian salt gland or elasmobranch rectal gland, both of which are comprised of relatively homogenous cell populations, the gills of bony fish are extremely complex in their histological makeup. To some extent, this complexity is due to the dual functions of gas exchange and ionic regulation that are served by this organ. The gill consists of two functionally specialized epithelia, that covering the respiratory lamellae and that surrounding the cartilaginous shaft and associated primary filamentous arteries. This latter epithelium, which in some fish species is duplicated on the inner surfaces of the opercular flaps that cover and protect the gill arches, contains the ion secreting chloride cells. Chloride cells are embedded in a complex stratified epithelium consisting of many cell types. They are unique in that they are the single cell type in the epithelium to both open to the environmental surface and to extend to the basal lamina. This fact alone suggested, shortly after the discovery that the gill was the primary site of extra-renal ion secretion in marine teleosts (Keys, 1931), that chloride cells were the active agents in this process (Keys and Willmer, 1932). Recent electrophysiological studies (Foskett and Scheffey, 1982) now have definitively identified the chloride cell as the primary site of salt secretion in the gill.

In stenohaline marine teleosts or in euryhaline teleosts adapted to seawater, chloride cells are numerous in the branchial epithelium. The cells are readily distinguished at the light microscopic level by their large size and by the cup-shaped depression or crypt that characterizes their apical surface (Philpott and Copeland, 1963). Invariably this crypt is shared by at least two chloride cells or by a chloride cell and one or more immature chloride cells which are termed accessory cells (Hootman and Philpott, 1980). In chloride cells, basolateral plasma membranes are amplified into an anastomosing network of tubules that penetrates almost the entire cell body (Philpott and Copeland, 1963; Pisam, 1981). Continuity of these tubular membranes with the cell surface was demonstrated by use of electron dense tracers (Ritch and Philpott, 1969). These weblike arrays of tubules surround and enmesh many mitochondria (Fig. 7). Both

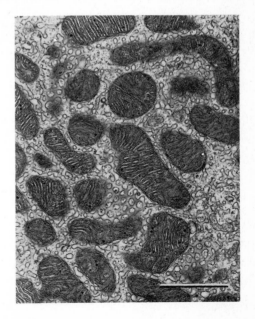

Figure 7. Cytoplasm of a chloride cell from the branchial epithelium of the pinfish Lagodon rhomboides. The pervasive nature of the labyrinth of plasma membrane tubules is well illustrated, as is the intimate association between these membranes and abundant mitochondria. Modification of the labyrinth into tightly anastomosed plaques is seen in several places. Scale bar is 1 μm.

mitochondria and basolateral membrane tubules are excluded from a
narrow zone beneath the apical cell surface, which is filled instead
with microfilaments and many vesicles, both coated and smooth surfaced
(Fig. 8). Under proper conditions of fixation, fusion of these vesicles

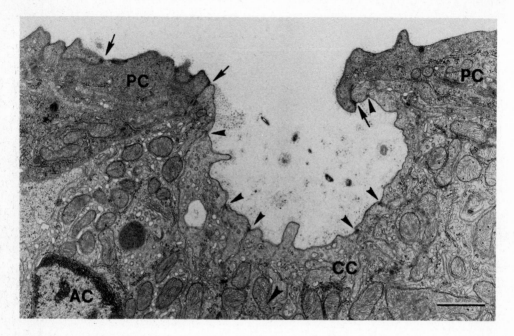

Figure 8. Apical crypt area of pinfish branchial epithelium. A chloride
cell (CC) and accessory cell (AC) share a common crypt bounded by pavement cells (PC). Note the presence of numerous coated and smooth vesicles
in the apex of the chloride cell. Occluding junctions between pavement
cells or pavement and accessory cells (arrows) are extensive, while those
that unite the accessory cell and chloride cell (arrowheads) are shallow.
Scale bar is 1 μm.

with the crypt plasma membrane is readily seen (Bradley, 1981). Electron
micrographs of thin sections of the apical chloride cell surface usually
show many profiles of occluding junction within a single crypt (Fig. 8).
Their presence results from the interdigitation of chloride cell and
accessory cell processes (Ernst, et al., 1980; Dunel-Erb and Laurent,
1980). Occluding junctions between chloride cells and accessory cells
differ markedly in their appearance from those between either of these
cell types and the surrounding pavement cells at the margins of crypts.
The former usually consist of two parallel strands, while the latter,
like those between neighboring pavement cells, are composed of a multi-
stranded meshwork (Fig. 9).

Figure 9. Apical crypt freeze-fractured replica from the opercular epithelium of the killifish Fundulus heteroclitus. Occluding junctions between chloride cell and accessory cell processes (arrowheads) consist of two parallel strands, while those between chloride cells and pavement cells (arrows) are composed of a net-like arrangement of strands. Numerous openings of the membrane labyrinth to the lateral chloride cell surface (P fracture face) are visible at lower right. Scale bar is 1 μm.

As in the avian salt gland, an increase in functional demand on the salt secreting epithelium of the gill results in a substantial increase in Na,K-ATPase activity (Forrest, et al., 1973; Hootman and Philpott, 1979). This increase coincides with proliferation and enlargement of chloride cells. This correlation suggested that chloride cells in marine teleosts would be rich in Na,K-ATPase activity and that the ATPase must play a central role in the ability of these cells to secrete salt. This initial suggestion was verified by cell isolation techniques, where we demonstrated that Na,K-ATPase activity was enriched in partially purified chloride cell suspensions (Hootman and Philpott, 1978), and by autoradiographic localization of ^{3}H-ouabain binding sites (Karnaky, et al., 1976; Ernst, et al., 1980). Using the Sr-NPPase procedure, we extended this localization to the ultrastructural level (Hootman and Philpott, 1979). The principal site of reaction product deposition within the branchial epithelium was the basolateral plasma membrane labyrinth of the chloride cell (Fig. 10). The less well developed labyrinth of accessory cells

displayed a much lower level of activity, while the membranes of pavement cells and of surrounding supportive and other cell types were almost
devoid of precipitates. An additional site of reaction product deposition

Figure 10. Sr-NPPase localization
of Na,K-ATPase activity in pinfish gill. Portions of three or
more chloride cells (CC) are seen
as well as basal (BC) and endothelial cells (EC). Precipitates
line the membranes of the chloride cell labyrinth, but are absent from basal cell and endothelial cell surfaces. Scale bar
is 2 μm.

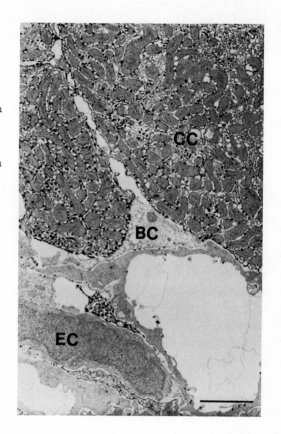

also was noted in chloride cells. Many of the vesicles in the rim of
cytoplasm bordering the apical crypt also exhibited significant levels of
activity, indicating that they may arise from the labyrinth tubules, a
possibility suggested previously on the basis of morphological observations (Bradley, 1981).

III. A model for hypertonic salt secretion

As can be seen from the preceding sections, salt secreting epithelia
in marine vertebrates share a number of strikingly similar morphological
and enzymatic features, despite the phylogentic diversity of the animals
in which they reside. Most notable among these are a highly amplified

basolateral cell surface rich in Na,K-ATPase activity, numerous mitochondria, extensive lateral folding of the cell surface, which increases
the linear amount of cell-to-cell edge contact, and simplified occluding
junctions. The fact that Na,K-ATPase in these secretory cells is oriented in such a way as to extrude Na^+ into lateral intercellular spaces,
coupled with the apparent modification of occluding junctions to potentiate paracellular flow of solutes, suggests that Na^+ enters the tubular
lumen in salt or rectal glands, or apical crypt in branchial/opercular
epithelium, by traversing the junction.

Earlier, we suggested that the movement of Na^+ across salt gland epithelium by this route could be supported by a transepithelial electrical
gradient (mucosal side negative) established by active transcellular Cl^-
transport (Ernst and Mills, 1977). Key features of this model were:
1) the presence of a Na^+/Cl^- co-transport mechanism in the basolateral
plasma membrane, 2) maintenance of high intracellular Cl^- activities,
and 3) the presence of an anion-selective channel in apical plasma membranes. In the model, Na,K-ATPase serves two functions, extrusion of Na^+
into lateral intercellular spaces and maintenance of the inwardly direct-
ed electrochemical gradient for Na^+ that indirectly energizes the coupled
uptake of Na^+ and Cl^-, the latter to levels high enough to potentiate its
diffusion across the apical cell surface. There is now clear evidence
that a small mucosa negative transepithelial potential difference is
characteristic of the secreting rectal gland (Silva, et al., 1977) and
opercular epithelium (Ernst, et al., 1980). Intracellular Cl^- concentrations of up to 80 mEq/l have been measured in the secretory cells of both
duck salt gland (Hootman and Ernst, 1981) and dogfish rectal gland (Silva,
et al., 1977), and evidence for a Na^+/Cl^- co-transport mechanism has been
obtained in all three epithelia (Degnan, et al., 1977; Hootman and Ernst,
1981; Ernst and Van Rossum, 1982; Hannafin, et al., 1983). Little definitive data is available presently regarding the functional attributes of
a putative apical membrane Cl^- channel, although its presence is indicated by the observation that intracellular Cl^- levels in both rectal gland
and salt gland cells fall abruptly when these cells are stimulated with
secretagogues (Silva, et al., 1977; Hootman and Ernst, 1981).

We have sketched in the broad outlines of the secretory mechanisms
operating in these epithelia and have identified some of the particular
transport elements involved. We have as yet, however, little knowledge
of the molecular characteristics of most of these elements or of the
ways in which their functional status is regulated as epithelial secretory activity is modulated by hormones and neurotransmitters. Biochemical
characterization of transport elements such as the Na^+/Cl^- cotransporter
and apical anion channels, identification of intracellular second messen-

gers in each epithelia, and elucidation of how these components of the secretory process interact at the molecular level to effect extrusion of NaCl are goals that must be vigorously pursued if we are to comprehend more fully how these specialized epithelia contribute to ionic homeostasis in marine vertebrates.

Acknowledgement

We wish to dedicate this paper to Drs. Frank P. Conte, Richard A. Ellis, and Charles W. Philpott, who have been our teachers, colleagues, and friends, and who have, through their enthusiasm for the study of epithelial transport processes, made possible our own contributions to this field.

References

Bonting SL, Caravaggio LL, Canady MR, and Hawkins NM (1964) Studies on sodium-potassium activated adenosinetriphosphatase XI. The salt gland of the herring gull. Arch. Biochem. Biophys. 106: 49-56.

Bradley TJ (1981) Improved visualization of apical vesicles in chloride cells of fish gills using an osmium quick-fix technique. J. Exp. Zool. 217: 185-198.

Bulger RE (1963) Fine structure of the rectal (salt-secreting) gland of the spiny dogfish Squalus acanthias. Anat. Rec. 147: 95-127.

Burger JW and Hess WN (1960) Function of the rectal gland in the spiny dogfish. Science 131: 670-671.

Dahl JL and Hokin LE (1974) The sodium-potassium adenosinetriphosphatase. Ann. Rev. Biochem. 43: 327-356.

Degnan KJ, Karnaky KJ Jr, and Zadunaisky JA (1977) Active chloride transport in the in vitro opercular skin of a teleost (Fundulus heteroclitus), a gill-like epithelium rich in chloride cells. J. Physiol. (Lond.) 271: 155-191.

Dunel-Erb S and Laurent P (1980) Ultrastructure of marine teleost gill epithelia: SEM and TEM study of the chloride cell apical membrane. J. Morphol. 165: 175-186.

Ellis RA, Goertemiller CC Jr, and Stetson DL (1977) Significance of extensive 'leaky' cell junctions in the avian salt gland. Nature 268: 555-556.

Ernst SA (1972a) Transport adenosine triphosphatase cytochemistry. I. Biochemical characterization of a cytochemical medium for the ultrastructural localization of ouabain-sensitive, potassium-dependent phosphatase activity in the avian salt gland. J. Histochem. Cytochem. 20: 13-22.

Ernst SA (1972b) Transport adenosine triphosphatase cytochemistry. II. Cytochemical localization of ouabain-sensitive, potassium-dependent phosphatase activity in the secretory epithelium of the avian salt gland. J. Histochem. Cytochem. 20: 23-38.

Ernst SA (1975) Transport ATPase cytochemistry: ultrastructural localization of potassium-dependent and potassium-independent phosphatase activities in the rat kidney cortex. J. Cell Biol. 66: 586-608.

Ernst SA, Dodson WC, and Karnaky KJ Jr (1980) Structural diversity of occluding junctions in the low-resistance chloride-secreting opercular epithelium of seawater-adapted killifish (Fundulus heteroclitus). J. Cell Biol. 87: 488-497.

Ernst SA and Ellis RA (1969) The development of surface specialization in the secretory epithelium of the avian salt gland in response to osmotic stress. J. Cell Biol. 40: 305-321.

Ernst SA, Goertemiller CC Jr, and Ellis RA (1967) The effect of salt regimens on the development of (Na^+-K^+)-dependent ATPase activity during the growth of salt glands of ducklings. Biochim. Biophys. Acta 135: 682-692.

Ernst SA and Hootman SR (1981) Microscopical methods for the localization of Na^+,K^+-ATPase. Histochem. J. 13: 397-418.

Ernst SA, Hootman SR, Schreiber JH, and Riddle CV (1981) Freeze-fracture and morphometric analysis of occluding junctions in rectal glands of elasmobranch fish. J. Membrane Biol. 58: 101-114.

Ernst SA and Mills JW (1977) Basolateral plasma membrane localization of ouabain-sensitive sodium transport sites in the secretory epithelium of the avian salt gland. J. Cell Biol. 75: 74-94.

Ernst SA and Mills JW (1980) Autoradiographic localization of tritiated ouabain-sensitive sodium pump sites in ion transporting epithelia. J. Histochem. Cytochem. 28: 72-77.

Ernst SA, Riddle CV, and Karnaky KJ Jr (1980) Relationship between localization of Na^+-K^+-ATPase, cellular fine structure, and reabsorptive and secretory electrolyte transport. In: Bronner F and Kleinzeller A (eds.) Current Topics in Membranes and Transport. vol. 13. Academic Press, New York, pp. 355-385.

Ernst SA and Schreiber JH (1981) Ultrastructural localization of Na^+,K^+-ATPase in rat and rabbit kidney medulla. J. Cell Biol. 91: 803-813.

Ernst SA and Van Rossum GDV (1982) Ions and energy metabolism in duck salt gland: possible role of furosemide-sensitive co-transport of sodium and chloride. J. Physiol. (Lond.) 325: 333-352.

Eveloff J, Karnaky KJ Jr, Silva P, Epstein FH, and Kinter WB (1979) Elasmobranch rectal gland cell: autoradiographic localization of ^{3}H-ouabain-sensitive Na,K-ATPase in rectal gland of dogfish, Squalus acanthias. J. Cell Biol. 83: 16-32.

Forrest JN Jr, Cohen AD, Schon DA, and Epstein FH (1973) Na transport and Na-K-ATPase in gills during adaptation to seawater: effects of cortisol. Am. J. Physiol. 224: 709-713.

Foskett JK and Scheffey C (1982) The chloride cell: definitive identification as the salt-secretory cell in teleosts. Science 215: 164-166.

Goertemiller CC Jr and Ellis RA (1976) Localization of ouabain-sensitive, potassium dependent nitrophenyl phosphatase in the rectal gland of the spiny dogfish, Squalus acanthias. Cell Tissue Res. 175: 101-112.

Hannafin J, Kinne-Saffran E, Friedman D, and Kinne R (1983) Presence of a sodium-potassium chloride cotransport system in the rectal gland of Squalus acanthias. J. Membrane Biol. 75: 73-83.

Hokin LE, Dahl JL, Deupree JD, Dixon JF, Hackney JF, and Perdue JF (1973) Studies on the characterization of the sodium-potassium transport adenosine triphosphatase. X. Purification of the enzyme from the rectal gland of Squalus acanthias. J. Biol. Chem. 248: 2593-2605.

Hootman SR and Ernst SA (1980) Dissociation of avian salt gland: separation procedures and characterization of dissociated cells. Am. J. Physiol. 238: C184-C195.

Hootman SR and Ernst SA (1981) Effect of methacholine on Na$^+$ pump activity and ion content of dispersed avian salt gland cells. Am. J. Physiol. 241: R77-R86.

Hootman SR and Philpott CW (1978) Rapid isolation of chloride cells from pinfish gills. Anat. Rec. 190: 687-702.

Hootman SR and Philpott CW (1979) Ultracytochemical localization of Na$^+$,K$^+$-activated ATPase in chloride cells from the gills of a euryhaline teleost. Anat. Rec. 193: 99-130.

Hootman SR and Philpott CW (1980) Accessory cells in teleost branchial epithelium. Am. J. Physiol. 238: R199-R206.

Karnaky KJ Jr, Kinter LB, Kinter WB, and Stirling CE (1976) Teleost chloride cell. II. Autoradiographic localization of gill Na,K-ATPase in killifish Fundulus heteroclitus adapted to low and high salinity environments. J. Cell Biol. 70: 157-177.

Keys AB (1931) Chloride and water secretion and absorption by the gills of the eel. Z. Vgl. Physiol. 15: 364-388.

Keys AB and Willmer EN (1932) Chloride secreting cells in the gills of fishes with special reference to the common eel. J. Physiol. (Lond.) 76: 368-378.

Kyte J (1976a) Immunoferritin determination of distribution of (Na$^+$+ K$^+$)ATPase over the plasma membranes of renal convoluted tubules. I. Distal segment. J. Cell Biol. 68: 287-303.

Kyte J (1976b) Immunoferritin determination of distribution of (Na$^+$+ K$^+$)ATPase over the plasma membranes of renal convoluted tubules. II. Proximal segments. J. Cell Biol. 68: 304-318.

Martin BJ and Philpott CW (1973) The adaptive response of the salt glands of adult mallard ducks to a salt water regime: an ultrastructural and tracer study. J. Exp. Zool. 186: 111-122.

Mayahara H, Fujimoto K, Ando T, and Ogawa K (1980) A new one-step method for the cytochemical localization of ouabain-sensitive, potassium-dependent p-nitrophenylphosphatase activity. Histochemistry 67: 125-138.

Mayahara H and Ogawa K (1980) Ultracytochemical localization of ouabain-sensitive, potassium-dependent p-nitrophenylphosphatase activity in the rat kidney. Acta Histochem. Cytochem. 13: 90-102.

Philpott CW and Copeland DE (1963) Fine structure of chloride cells from three species of Fundulus. J. Cell Biol. 18: 389-404.

Pisam M (1981) Membranous systems in the chloride cell of teleostean fish gill; their modifications in response to the salinity of the environment. Anat. Rec. 200: 401-414.

Riddle CV and Ernst SA (1979) Structural simplicity of the zonula occludens in the electrolyte secreting epithelium of the avian salt gland. J. Membrane Biol. 45: 21-35.

Ritch R and Philpott CW (1969) Repeating particles associated with an electrolyte-transporting membrane. Exp. Cell Res. 55: 17-24.

Schwartz M, Ernst SA, Seigel GJ, and Agranoff BW (1981) Immunocytochemical localization of (Na$^+$,K$^+$)-ATPase in the goldfish optic nerve. J. Neurochem. 36: 107-115.

Silva P, Stoff J, Field M, Fine L, Forrest JN, and Epstein FH (1977) Mechanism of active chloride secretion by shark rectal gland: role of Na-K-ATPase in chloride transport. Am. J. Physiol. 233: F298-F306

Stewart DJ, Semple EW, Swart GT, and Sen AK (1976) Induction of the catalytic protein of (Na$^+$+K$^+$)-ATPase in the salt gland of the duck. Biochim. Biophys. Acta 419: 150-163.

Stirling CE (1976) High resolution autoradiography of ^{3}H-ouabain binding in salt transporting epithelia. J. Microscopy 106: 145-157.

MODELS OF SALT AND WATER FLOW ACROSS EPITHELIA : AN EVALUATION BY ELECTRON PROBE X-RAY MICROANALYSIS

B.L. GUPTA

I. Epithelial transport: an overview

It has been appreciated for a long time (Townson, 1799) that epithelia such as skins and urinary bladders of certain amphibia and reptiles are important in physiological regulation of salt and water. In order to obtain more satisfactory physiological data and to determine the microscopic basis of salt and water transport, experimental preparations of isolated epithelia were introduced early in the last century. The chamber (Matteucci and Cima, 1845) designed to examine the osmotic flow across the flat-sheet preparations of frog skin, bovine urinary bladder etc bears a close resemblance to the Ussing-type chambers, the current variations of which are reviewed by Lahlou in the present Symposium. The progress in our understanding of epithelial transport phenomena over the past 140 years has followed the development of new techniques, many of which have been recently reviewed by Macknight and Leaf (1978), Harvey (1982) and Civan (1983). This experimental work has resulted in a phenomenological description of ions and water transport in epithelia which is summarised in Fig. 1 and in a plethora of reductionistic (Diamond, 1982), macroscopic (Maddox, 1983) or thermodynamic 'black box' models (Gupta et al. 1977), two of which are discussed below. Unfortunately, this reductionistic approach, permissible and in fact necessary to explain natural phenomena (Maddox, 1983), has been extended to the experimental preparations as well, so that the epithelial tissues in such studies bear lesser and lesser resemblance to their structural organization in the whole organism. Thus for

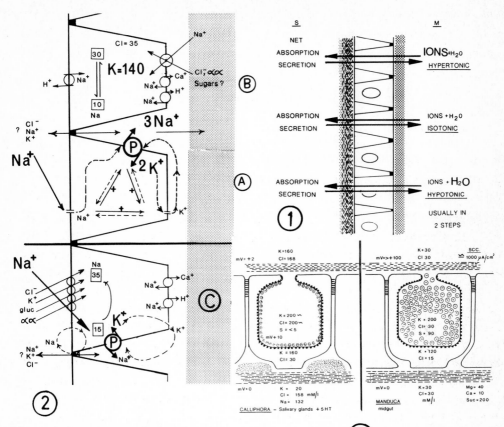

Figs. 1-3. For explanation see text.

③

'experimental convenience' the vertebrate gall bladders are
investigated in 50 mosM salines (Zeuthen, 1983) or even in tap-water
(Hill and Hill, 1978) while the energy consuming osmoregulatory
mechanisms in these animals have evolved to maintain the tissue and
plasma body fluid concentrations within a very narrow range! The
serosal and other connective tissue elements are completely and
throroughly stripped off the gall bladder to obtain 'optically clean'
preparations (Coble et al. 1982) and from frog skin ('split
preparation') for electrophysiological measurements (see Civan,
1983). Such preparations do indeed provide experimentally elegant
information but one wonders how far is that knowledge directly
applicable to the understanding of transport functions under 'messy'
conditions in vivo! Or how far can the data from such 'minimum'
preparations be compared with results obtained from complete tissues
in vitro and in vivo! A case in point is the critical role of
extracellular mucus and other similar matrices in the intestinal

tissues of vertebrates (Smithson et al. 1981; Kirsch et al. this Symposium) and of insects (Dow et al. 1981, 1983) in maintaining a relatively constant ionic environment next to the membranes of the transporting cells. In this respect the invertebrate epithelia such as Malpighian tubules, salivary glands, nephridia, gut etc have a great advantage as experimental preparations (Gupta and Hall, 1983) to study transport.

The current era for the study of ion and water transport was ushered in when Ussing and Zerahn (1951) introduced short circuited preparation of frog skin to rigorously define "active" absorption of Na^+. Skou (1957) discovered a Na^+, K^+-activated adenosine triphosphatase (Na^+, K^+-ATPase) bound to cell membranes. This led to the proposal by Koefoed-Johnsen and Ussing (1958) of a 'black box' model (KJU) for transepithelial Na-transport which has since dominated the thinking in transport physiology (Diamond, 1982; Civan, 1983). After the realisation that the transport of water is coupled to the net transport of solute (often NaCl), Curran (1960) adapted KJU model into a 'double-membrane' theory to explain isotonic absorption in rat intestine. The efforts to seek detailed 'microscopic' or 'quantitative' models in order to accommodate tissue fine structure (see Skadhauge in Gupta et al. 1977) have resulted in an ongoing debate in which the validity of the basic models has often been questioned. However, it now seems that much of this scepticism of the basic models was based on either faulty or incomplete information obtained by inadequate techniques previously available to measure critical parameters such as (i) the concentrations and activities of ions in the transporting cells (Civan, 1983; Diamond, 1982), (ii) permeabilities of cell membranes (Fischbarg and Montoreano, 1982) and cell junctions (Bradley and Purcell, 1982), and (iii) the exact composition and osmotic concentration of fluid in the relevant intercellular spaces involved in transepithelial water flows (Gupta and Hall, 1979, 1981). Here we briefly survey the contributions made by recent studies with the electron probe x-ray microanalysis (EPXMA) and ion selective microelectrodes in resolving some of these issues. The methodology of EPXMA has recently been reviewed by Gupta and Hall (1982) and Hall and Gupta (1983) and that of ion-selective microelectrodes by Civan (1983): these need not be described here.

II. KJU model and intracellular ions

The KJU model was based on the study of isolated, short circuited frog skin bathed for baseline control in amphibian Ringer on both sides. It postulated that for ion transport this complex multilayered transitional epithelium can be modeled as a 'black box' syncytium with two membranes (Fig. 2A). The outer, apical or mucosal membrane normally facing the pond water behaves as a Na-electrode and allows Na^+ to enter the cells down its electrochemical activity gradient. The inner, basal or serosal membrane of the 'box' facing the blood plasma is permeable to K^+ and has the Na^+/K^+ pump which extrudes this Na^+ in exchange for K^+ and maintains a low Na^+ high K^+ concentration inside the box (cells). The whole mechanism then results in a net transepithelial influx of NaCl even from very low Na^+ concentrations (<2mM) normally found in the outside medium such as pond water. Similar experimental results from a simple epithelium lining the urinary bladder of toads extended the validity of the KJU model (see Civan 1983). Discovery of gap-junctions permitting movement of ions and small molecules from cell to cell even in transitional or multilayered epithelia such as frog skin (see Civan, 1983) has now vindicated the assumption of syncytium in the original KJU model. Similarly, Na^+, K^+-ATPase has with some notable exceptions (Gupta and Hall, 1983) now been localised on the basolateral-membranes of not only the frog skin but in the ion transporting epithelia in a range of animals from different Phyla, irrespective of the direction of transport in Fig. 1 (see Hootman SR in this Symposium). This simple KJU model has therefore been adapted to explain active Na^+ transport and Na^+-linked secondary transport (Fig. 2B,C) (Diamond, 1982).

For the KJU model to operate successfully, it was necessary to show that the intracellular concentrations and activities of Na^+ in various epithelia are indeed low enough to allow a downhill entry of Na^+ from the apical side. However, right up to the late 1970's even in epithelia with physiologically tight junctions (high electrical resistance) such as frog skin and toad bladder, the chemical estimates of cell-Na continued to yield values in the range of 50 mM while the estimates of ^{24}Na transport pool from the mucosal side were only about 10 mM (Macknight and Leaf, 1978; Civan, 1983). Moreover, under various experimental conditions where the mucosal and/or serosal bath was altered, the changes in intracellular ions and transport pools did not seem to be consistent with the KJU model (references in Civan, 1983). Measurements with ion-selective microelectrodes did not provide unequivocal results in different laboratories: the microelectrode values for intracellular Na^+ and K^+ in frog skin and toad bladder still remain uncertain (Civan, 1983; DeLong and Civan, 1983). For example the microelectrode estimates for cell-K^+ have ranged from 40 mM to 180 mM over the past 20 years (Civan, 1983). Furthermore, in spite of some heroic attempts (Zeuthen and Monge, 1975) the use of ion-selective microelectrodes to simultaneously measure Na^+, K^+, Cl^- in the same cell have not been successful.

On the other hand EPXMA with energy dispersive (Si-Li) spectrometers (EDS) provides simultaneously a complete spectrum of all the elements with atomic numbers $Z \geq 11$ (in some current models $Z \geq 4$) and also permits the estimates of local dry mass and hence water at every analysed site under the electron beam (Gupta and Hall, 1982). The data on intracellular concentrations of Na, K, Cl under baseline conditions in a variety of transporting epithelia in animals are summarised in Table 1. For frog skin the baseline value of 9 mmol/kg wet wt (=12 mmol/l H_2O) is consistent with the previous estimate of 10 mM for the ^{24}Na transport pool from the mucosal side. Rick et al. (1978a) have also shown that with zero-Na on the mucosal side the cell-Na drops nearly to zero but the replacement of Na with choline on the serosal side does not significantly change cell-Na levels. Similarly, amiloride added to the mucosal bath reduces the

cell-Na to 7 mmol/l H_2O. Similar results have also been obtained by Rick et al. (1978b) for the urinary bladder of the toad Bufo marinus using frozen dried cryosections and EDS. These results have been confirmed by Civan et al. (1980) in toad hemibladder under open circuit conditions analysed both in frozen-hydrated and frozen-dried cryosections using EDS for all elements and wave length dispersive spectrometer (WDS) for Na. As discussed in detail by Hall and Gupta (1983) and Civan (1983) WDS offer a distinct advantage over EDS for measuring low levels of Na (20 mmol/kg wet or dry mass). It is noteworthy that the value of 18 mmol/kg wet wt for cell-Na in toad bladder obtained by Civan et al. is some 50% higher than the value obtained by Rick et al. but is similar to a recent estimate of 18 mmol/l H_2O measured in frog skin with Na^+-selective microelectrodes (see Civan, 1983). These EPXMA studies therefore remove a long standing anomaly between estimates for the total cell-Na and the Na-transport pools and also eliminate the need to postulate that some 80% of the cell-Na is bound or compartmentalised e.g. in nuclei (see Gupta and Hall, 1982). [In fact the comparison of EPXMA values for the total Na, K, Cl in cells with the microelectrode values for the

Table 1. Intracellular concentrations of Na, K, Cl in transporting epithelia
under baseline state measured by x-ray microanalysis

TISSUE	mmol/kg wet weight			REFERENCE
	Na	K	Cl	
INSECTS				
Calliphora Malpighian tubules	26	143	–	Gupta & Hall (1974)[a]
Rhodnius Mal.tub. in vitro	12	124	45	Gupta et al. (1976)
Cricket Mal.tub. in vivo	–	101	41	Marshall (1980)
Calliphora salivary gland in vitro	15	125	23	Gupta et al. (1978a)
Calliphora rectal papillae in vivo	23	85	28	Gupta et al. (1980)
Periplaneta sal.gl. in vitro				
fluid sec. (P) cells	11	123	38	Gupta & Hall (1983)
zymogen (C) cells	11	142	39	Gupta & Hall unpubl.
Schistocerca gut caeca				Dow et al. (1981)
In vivo, epith. cell	11	133	25	
Manduca larval midgut				Dow et al. (1983)
In vitro, short circuit				
goblet cells	0	130	21	
columnar cells	0	129	12	
In Vivo, goblet cells	5	91	16	
columnar cells	5	127	19	
VERTEBRATES				
Frogskin, in vitro, short circuit	9	118	35	Rick et al. (1978a)
Toad, urinary hemibladder				
short circuit, granular cells	12	111	32	Rick et al. (1978b)
open circuit, granular cells	18	116	28	Civan et al. (1980)
Rabbit ileum, enterocyte in vitro	15	110	31	Gupta et al. unpubl.
Rat kidney, proximal tubules	21	140	27	Beck et al. (1980)
distal tubule	12	140	14	
Duckling salt gland, epith. cells				Andrews et al. (1983)
salt water adapted	23	118	36	
fresh water adapted	24	113	40	
CNIDARIA				
Rhodactis, sea-anemone				Lubbock et al. (1981)
nematocyte, cytoplasm	43	108	106	
columnar cells, cytoplasm	32	98	101	

a Cited by Gupta (1976)

activities suggests an intracellular activity coefficient of 0.7 to
0.5 for these ions.]

Table 1 shows that in a variety of epithelia not only is the
cell-Na low but the cell-K is high and ranges between 100-150 mmol/kg
wet wt as predicted from the KJU model. In Na-transporting systems
high cell-K is maintained by the basolateral Na^+, K^+-pump. If under
experimental conditions serosal-K^+ is reduced to zero, one would
expect the inhibition of the pump, the abolition of the mucosal to
serosal Na-flux (SCC) and the loss of cell-K^+. Restoration of
serosal-K^+ should rapidly restore this function. However zero
serosal-K^+ experiments on toad bladder have in the past produced
results which did not seem consistent with the KJU model and needed
to postulate a second K^+-pumping mechanism not coupled to the mucosal
Na-entry (see Civan, 1983 for references). However the EPXMA of toad
bladder under zero serosal-K^+ conditions (Civan et al. 1980) has
shown that near normal levels of cell-K are maintained by the
continued activity of Na^+, K^+-pump with K^+ (and Na^+!) recycling along

the basolateral front. Inhibition of Na^+, K^+-pump with ouabain under
zero-K condition does indeed lead to some 80 mmol loss of cell-K and
a similar gain in cell-Na (from the serosal side) without much change
in cell-Cl.

Evidence for K-recycling (implied in the KJU-model in Fig 2A)
has since been reinforced by microelectrode studies both in toad
bladder and frog skin (DeLong and Civan, 1983; Civan, 1983). Indeed
ion-recycling both across the pump-leak pathways on the basolateral
side and paracellular-mucosal pathways on the apical side, together
with the conservation of ions in the immediate vicinity of cell
membranes by charged matrices, plays a very important part in
transepithelial transport. Such recycling of ions is not generally
evident in flux measurements and in electrophysiological parameters.
These factors are frequently overlooked in the design and
interpretation of experiments. As noted by Diamond (1982) and
discussed by Civan (1983) at some length, the results from many
previous experiments substantially deviating from baseline conditions
have often been interpreted too rigidly without allowance for changes
in ionic and hydraulic conductances of cell membranes effecting
transcellular pathways and of intercellular junctions for
paracellular pathways. More critical examination of these mechanisms
has added many details and refinements to the 'black-box' model of
Koefoed-Johnsen Ussing which are summarised in Fig. 2. The upper
section of the figure refers to tight epithelia and the lower section

to leaky epithelia (intestine). References and discussions for many of these refinements are to be found in Diamond (1982), Civan (1983) and Bradley and Purcell (1982).

A. Active transport of Chloride

Chloride transport appears to play a particularly significant role in marine and brackish water adaptations as is reflected in the proceedings of this Symposium. In most epithelia working with a basolateral Na^+, K^+-pump the intracellular levels of Cl^- are several fold higher than expected from a simple Nernstian distribution. Cl^- values similar to those in Table 1 have also been found with Cl^--selective microelectrodes in several other vertebrate epithelia (Frizzel and Duffey, 1980). Such high concentrations of cell-Cl are generally explained either by an uphill entry of Cl^- on a Na^+, Cl^- symtransport and/or a Cl^-/HCO_3 exchange as a part of the cell-pH regulation. However a primary active mechanism of Cl^--transport has been proposed in many epithelia (Harvey, 1982). The presence of HCO_3 and Cl^--dependent ATPase in the rectum of larval dragonflies has been claimed (Komnick et al. 1980) but remains to be proved.

B. Intracellular Na^+ and stimulation of transport by hormones and nutrients.

Many physiological agents such as antidiuretic hormones, aldosterone, luminal nutrients (sugars, amino acids) increase Na^+ entry into the cells and also stimulate transepithelial transport. The evidence for an expected rise in cell-Na^+ by chemical estimates and microelectrode measurements has been equivocal (Diamond, 1982) and therefore the mechanism for the stimulation of Na^+, K^+-pump has remained unclear. Table 2 summarises the EPXMA data on four diverse epithelia and in each case after stimulation the cell-Na has increased about 3-fold. [Similar increases in cell-Na have now been reported in a range of other phenomena such as stimulation of exocytotic secretion, mitogenesis, fertilization, normal or malignant cell proliferation etc.] At least in a mammalian kidney preparation the kinetic studies have shown that a rise in cell-Na^+ from 10 mM to 37 mM will alone increase the activity of basolateral Na^+, K^+-ATPase from 5% to 100% - a 20-fold stimulation in pumping function. A rise in cell-Na^+ could well be the major if not the only mechanism by which various physiological agents stimulate transepithelial transport (Gupta and

Table 2. Changes in cell-Na in cell stimulation measured by x-ray microanalysis

TISSUE IN VITRO	mmol/kg wet wt			TARGET STIMULATED	REFERENCE
	Na	K	Cl		
RHODNIUS, Mal. tubule					Gupta et al. (1976)
baseline control	12	124	45	apical cation	
+ 5-HT	37	126	45	(K)-pump	
COCKROACH, sal. gland, fluid (P) cells					Gupta and Hall (1983)
baseline control	11	123	38	apical Na/K	
+ dopamine; 5-HT	25	158	29	pump?	
FROGSKIN, short circuit					Rick et al. (1983)
baseline control	12	118	35	basilateral	
+ vasopressin	29	101	36	Na/K pump	
RABBIT, enterocyte					Gupta et al. (1978 and
baseline control	15	135	31	basilateral	unpublished)
+ mucosal sugar	36	115	50	Na/K pump	
SEA ANEMONE, nematocyte					Lubbock et al. (1981)
resting control	43	108	106	nematocyst	
electric shock	117	116	167	discharge	

Table 3. Estimates of interspace hypertonicity in some fluid transporting epithelia.

METHOD	TISSUE	EMERGENT FLUID	INTERSP. EXCESS	WORKERS
Streaming potentials	Rabbit gall bladder	Isotonic	20 mM	Machen & Diamond (1969)
Micropuncture/ freezing point	Cockroach rectal pads	hypotonic	130 mosM	Wall et al. (1970)
EPXMA	Blowfly rectal papillae	hypotonic	155 mM	Gupta et al. (1980)
EPXMA	Rabbit ileum	Isotonic	20-40 mM	Gupta & Hall (1981)
Microelectrodes	Necturus gall bladder	Isotonic	< 18 mM	Simon et al. (1980)
Interferometry	Necturus gall bladder	Isotonic	<20 mM	Coble et al. (1982)
Computer Models				
Osmosis	Various	isotonic	few mosM	Several
Electro-osmosis	gall bladder	Isotonic	zero	Hill (1975)

EPXMA = Electron probe x-ray microanalysis of frozen-hydrated sections.

Hall, 1983; Rick et al. 1983).

C. K$^+$-pumps in insects and extracellular matrices

Unlike other animals, the epithelial transport in many insects is dominated by an apically located ouabain-insensitive K$^+$-pump (Harvey 1982, but see Gupta and Hall, 1983). In phytophagous insects such as the larvae of many lepidoptera there is little Na$^+$ in the diet and the Na$^+$-levels in cells and haemolymph tend to be extremely low (Dow et al. 1981, 1983). Clearly the transport in such insect epithelia cannot be modeled strictly by KJU formalisation. However the presence of a basolateral Na$^+$, K$^+$-ATPase has been demonstrated even in phytophagus insects (grass-hoppers) and has not been rigorously ruled out in other species. Even when the apical K$^+$-pump dominates transport in insect epithelia, a ouabain sensitive Na$^+$, K$^+$-ATPase may still be involved under baseline conditions and in cell-volume regulation (see Gupta and Hall, 1983 for further discussion).

For comparative physiologists it is of particular interest to compare two different systems of insect transport with similar tissue geometry and an apical K$^+$-pump (Fig. 3). In <u>Calliphora</u> salivary glands an electrogenic apical K$^+$-pump secretes K$^+$ into a cavity (canaliculus): Cl$^-$ and H$_2$O follow to produce an isotonic fluid (Gupta and Hall, 1979). The transepithelial potential (TEP) is very low. If however the apical K$^+$-pump is stimulated with cAMP without an increase in Cl$^-$ conductance the TEP can rise to >200 mV but cannot be sustained. In <u>Manduca</u> larval midgut a similar K$^+$-pump maintains some 200 mM K$^+$ in the goblet cavity with a -vely charge matrix (about 30% dry wt) but this K$^+$-secretion is not followed by Cl$^-$ (or any other anion). This mechanism now sustains a TEP of >100 mV lumen +ve which is used to attain a pH 10 to 12 in the gut lumen and also to absorb nutrients by the columnar cells using the electrochemical gradient of K$^+$ instead of Na$^+$. A system very similar to the lepidopteran goblet cells with an extracellular Donnan-matrix is also present in most insect sensilla and apparently even in the chochlea of mammalian ear (see further discussion in Dow et al. 1983). It is possible that a similar matrix with -ve charges might be present in the cavity of Cl-cells (Zadunaisky in this Symposium) and maintains a low Cl$^-$ activity immediately outside the apical membrane, thus eliminating the need for a Cl$^-$-pump. Such -vely charged glycosaminoglycans are lost from conventional preparations for light and electron microscopy and require cryosections of unfixed frozen tissues for demonstration (Kirsch et al. in this Symposium).

III. Models for coupled transport of ions and water: interspace hypertonicity

The 'double-membrane' model (Fig. 4) of Curran (1960) was proposed to explain the absorption of an isotonic fluid even against

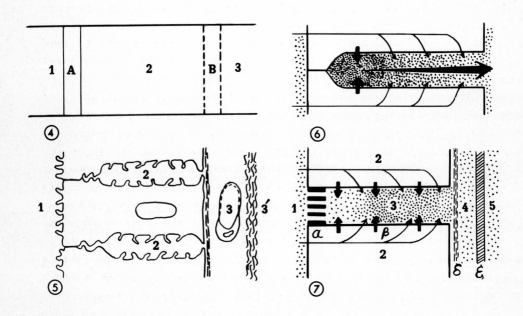

Fig. 4-7. Models for fluid tranpsort by local osmosis (see text for further explanation).

a transepithelial gradient of some 50 mosM in the small intestine of rat in vitro. It postulated that a selectively permeable barrier 'A' actively transports a solute (Na^+) from the compartment '1' (mucosal) into a compartment '2' within the tissue separated by a semipermeable barrier 'B' from the compartment '3' (serosal). The high concentration of solute (generally NaCl) thus generated in compartment '2' is osmotically equilibrated by drawing water from the compartment '1' and the resulting hydrostatic pressure due to restricted outflow through the barrier 'B' then pushes the isotonic absorbate into compartment '3'. [In reality the emergent fluid in isotonic transport in vitro is always 1-3% hypertonic.] On the evidence of tisssue fine structure and histochemical localisation of

(Na^+, K^+)-ATPase on the basolateral side, Curran's 'black-box' model was fitted to rabbit gall-bladder by Kaye et al. (1966) as summarised in Fig. 5 and to Calliphora rectal papillae by Gupta and Berridge (1966; see Gupta et al. 1980 for full references). The barrier 'B' in Kaye et al.'s interpretation is represented (a) by the narrow (0.1 μm) serosal opening of lis, and (b) by the basement lamina and serosal connective tissue composed of collagen and glycosaminoglycan with a high density of negative charges. It should be emphasised that in all fine structural studies thus far this serosal opening always appears highly restricted, irrespective of the extent of dilation of lis. This observation has been confirmed in rabbit ileum in frozen-hydrated and frozen-dried cryosections (STEM images in Hall and Gupta, 1983).

From the fine structural evidence of changes in the geometry of lis in rabbit gall bladder under different physiological states of fluid transport in vitro (now confirmed in the living tissue by Spring and Hope, 1978) Diamond and Bossert (1967) then constructed a quantitative (computer) model for solute-linked water transport (Fig. 6). In this model the role of the barrier 'B' is completely discarded and the osmotic equilibration of the fluid is explained by a 'standing gradient' of solute in lis. The ratio of length and radius of the lis and the hydraulic conductivity (Lp) of the cell membrane then determines the final osmolarity of the emergent fluid (hyperosmotic or isosmotic but never hypo-osmotic). In this model all the transepithelial flows were assumed to follow a transcellular route. The discovery that in most vertebrate epithelia transporting isotonic fluid the 'tight junctions' tend to be leaky (low electrical resistance) and therefore constitute paracellular routes of solute and water flow has complicated the issue. To accommodate such paracellular flows but also to reintroduce the function of serosal tissue, Sackin and Boulpaep (1975) proposed a more elaborate model of local osmosis for Necturus proximal tubule (Fig. 7). Sackin and Boulpaep pointed out that the salt reflective coefficient and restrictive properties of the barrier 'B' are critical in determining the level of lis hypertonicity that can be accommodated in generating an isotonic absorption by 'local osmosis'. Similarly, the radius of 'tight junction' channel and its reflection coefficient to ions also affect the degree of interspace hypertonicity (see Fischbarg in Bradley and Purcell, 1982).

Hill (1975, and in Gupta et al. 1977) has criticised all models for local and simple osmosis for salt-linked isotonic transport primarily because such models require the relevant cell-membranes to have osmotic permeabilities (Pos) and hydraulic conductivities (Lp) some 3 or 4 orders of magnitude higher than had been previously known. However more recent evidence from the osmotic behaviour of living tissues examined by the elegant light optical interferometric methods show that Lp and Pos values in many fluid transporting epithelia are indeed high (Persson and Spring, 1982; Fischbarg and Montoreano, 1982), thus reaffirming support for interspace hypertonicity models for local osmosis.

A major factor in much of the debate about the models of solute-linked fluid transport has been the lack of methods to directly measure the osmotic concentration of interspace fluid in situ. The only actual measurement of interspace fluid until 1976 was by Wall et al. (1970) on the micropuncture samples from the rectal pads of cockroach. In these experiments the tissue was removing water (hypotonic transport) from the lumen contents of some 800 mosM into the haemolymph of about 350 mosM. [As discussed elsewhere (Gupta, 1976; Gupta and Hall, 1983) such hypotonic transport is always a 2-step process achieved by transporting a primary isotonic fluid and subsequently removing solute from it by hypertonic transport.] Wall et al's value of 130 mosM for interspace hypertonicity has now been confirmed by in situ measurements with EPXMA in Calliphora rectal papillae (Table 3). These high values while consistent with the local osmosis cannot however be compared in general with the interspace hypertonicity values (measured or calculated) in isotonically transporting leaky epithelia of vertebrates and least with the values from tissues made to transport in a bath osmolarity of about 50 mosM (Zeuthen, 1983). The classic observation of Wall et al (1970) has been frequently misquoted (e.g. Zeuthen, 1983) and quite unjustifiably suspected (e.g. Hill, 1975; Coble et al. 1982).

Since 1976 Gupta and coworkers using EPXMA of hydrated cryosections have shown that the concentration of major solutes Na, K, Cl is indeed higher in the relevant interspaces of a variety of animal epithelia (full data in Gupta and Hall, 1982). Apart from the question of clearly resolving the free fluid in very narrow and complex interspaces (Gupta and Hall, 1979; 1981) EPXMA measures total concentration of each element in mmol/kg wet wt. Conversion of such EPXMA data into actual osmotic pressure of the fluid requires information not only on the interspace water contents (which can be measured) but also the extent of ion-binding to the -ve charges in glycosaminoglycans ('glycocalyx': see Revel JP in Bradley and

Purcell, 1982) and the exact boundary conditions (Lp, Pos etc) of interspaces. The last two factors would also apply to the microelectrode measurements (Simon et al., 1981; Zeuthen, 1983) where ion concentrations have to be calculated from tip potentials and need assumptions about the activity and osmotic coefficients of ions in the interspace fluid. Table 3 compares the values of lis hypertonicity in several epithelia measured by different methods. For vertebrate tissues transporting isotonically under normal bath osmolarities the interspaces appear to have about 20 mM more ions than in the bath. For reasons mentioned above, it is not easy to translate these measured values into exact osmotic pressure of the lis fluid within the tissue but it would be surprising if such excess of ions does not result in any hypertonicity at all as required by electro-osmosis (Hill, 1975) or paracellular transport through the junctions (Hill in Ussing et al. 1981). It is frequently overlooked by vertebrate physiologists that many insect epithelia (Section IIC) when bathed in normal Na-Ringer secrete isotonic KCl or Na+KCl at very high rates (Gupta and Hall 1979, 1981). Any universal model which proposes that most of the solute and water is dumped into interspaces as bulk-flow through cell junctions etc. cannot possibly apply to such epithelia.

IV. Ionic gradients in cells and Na^+,K^+-ATPase

In the 'standing gradient' model Diamond and Bossert (1977) postulated that the solute (Na) transport into lis was confined to the 'blind' end. An apparently uniform distribution of Na^+, K^+-ATPase along the basolateral membrane shown by histochemical techniques was thought to have made this unlikely. EPXMA studies now show that in sugar-stimulated enterocytes of rabbit ileum (Gupta et al. 1978b and unpublished data) and in vasopressin stimulated frog skin (Rick et al. 1983) the cell-Na shows a clear gradient from ~ 45 mmol near the mucosal side to about 10 mmol near the basal side (Fig. 8). A similar gradient of Na^+-activity in rabbit enterocytes in vivo measured with ion-selective microelectrodes was reported by Zeuthen and Monge (1975). Under baseline conditions of slow transport the cell-Na^+ gradient is much smaller (Fig. 8) and perhaps statistically insignificant. As noted in section IIB, for maximal activity the Na^+, K^+-ATPase requires about 37 mM cell-Na^+. A large mucosal to serosal Na^+ gradient in the cells would mean that most of the Na^+ secretion into lis would be confined to the apical part but the H_2O

input is likely to be distributed uniformly over the entire length of
lis. A gradient of Na, K and Cl was observed in the lis of sugar-

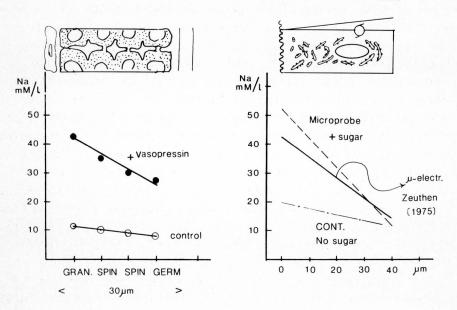

Fig. 8. Intracellular gradients of Na (left) in frogskin (Rick et
al. 1983); (right) rabbit enterocytes (Gupta et al. 1978b and Unpub;
Zeuthen & Monge, 1975).

stimulated rabbit ileum by Gupta et al (1978b). Osmotic gradients in
epithelial cells separating very large differences in osmotic
concentrations between mucosal and serosal sides have also been
postulated (Kirschner in Gupta et al. 1977) but remain to be proved.
Such large intracellular gradients of ions might explain why the
intracellular measurements from ion-selective microelectrodes (which
monitor tip potential at an undetermined and probably different
position in the cell at every impalement) frequently show such a
large spread. A gradient of Na^+, K^+-ATPase activity also means that
the membrane area for H_2O flow might well be several fold of the area
for Na^+ input (cf Durbin and Hill, 1981): this would assist in the
osmotic equilibration of the emergent fluid.

V. Conclusions

The measurements of intracellular Na, K, Cl with EPXMA and ion-
selective microelectrodes in several animal epithelia over the past

ten years have provided results entirely consistent with the Koefoed-Johnsen and Ussing model (1958) to explain an active transepithelial transport of Na^+. Generally low values of cell-Na^+ (<20 mM) in non-marine animal epithelia under baseline conditions and the rise in cell-Na^+ to some 40 mM after hormonal and other stimulation of transepithelial transport are also consistent with the electrochemical gradient of Na^+ as the energy source for the 'secondary active' transport of several other solutes. The KJU model has therefore provided the conceptual basis to explain sugar stimulation of Na^+-linked fluid absorption in human small intestine which has been proclaimed as '...potentially the most important medical advance this century' (Lancet, August 5, 1978; p.300) and has helped to revive the oral therapy of the osmoregulatory disease cholera - an ecological consequence of marine-terrestrial interface (Science American , 249(1): 21). In other animal epithelia which primarily rely on an active secretion of K^+ or Cl^-, the KJU principle of ionic gradients across two cell-membranes still holds.

For transepithelial fluid transport the EPXMA (and other) measurements support the hypertonic interspace models for local osmosis first postulated as a 'black-box' double-membrane concept by Curran (1960). Computer generated quantitative elaborations of osmosis models are usually constructed to explain transport in a specific epithelium. The validity of interspace hypertonicity values predicted by such models will depend on the exact transport parameters of all the interspace boundaries. Such boundary conditions are far from well-understood in any ion and water transporting epithelium (see Concluding Discussion in Bradley and Purcell, 1982) and must vary according to the structural and functional complexity of the epithelium. It might be futile therefore to seek a universal, fully quantitative (microscopic) model with a single value for interspace hypertonicity for all fluid transporting epithelia.

We are very grateful to Mrs J Schreiber for her patient effort in preparing the camera-ready copy of this manuscript and to Mr M Day for the photographic assistance.

References

Andrews SB Mazurkiewicz JE and Kirk RG (1983) The distribution of intracellular ions in the avian salt gland. J Cell Biol 96: 1389-1399.

Beck F Bauer R Bauer U Mason J Dorge A Rick R and Thurau K (1980) Electron microprobe analysis of intracellular elements in the rat kidney. Kidney Intl 17: 756-763.

Bradley SE and Purcell EF (1982) The Paracellular Pathway. Macy, New York.

Civan MM (1983) Epithelial ions and transport. Wiley-Interscience, New York Chichester Brisbane Toronto Singapore.

Civan MM Hall TA and Gupta BL (1980) Microprobe study of toad urinary bladder in absence of serosal K^+. J Membr Biol 55: 187-202.

Coble AJ Leader JP and Spring KR (1982) Microscope interferometry of Necturus gallbladder epithelium. In: Bradley and Purcell (1982), p.270.

Curran PF (1960) Na, Cl, and water transport by rat ileum in vitro. J Gen Physiol 43: 1137-1148.

DeLong J and Civan MM (1983) Microelectrode study of K^+ accumulation by tight epithelia: I. Baseline values of split frog skin and toad urinary bladder. J Membr Biol 72: 183-193.

Diamond JM (1982) Transcellular cross-talk between epithelial cell membranes. Nature (London) 300: 683-685.

Diamond JM and Bossert WH (1967) Standing gradient osmotic flow. A mechanism for coupling of water and solute transport in epithelia. J Gen Physiol 50: 2061-2083.

Dow JAT Gupta BL and Hall TA (1981) Microprobe measurements of Na, K, Cl, P, S, Ca, Mg, and H_2O in frozen-hydrated sections of anterior caeca of the locust Schistocerca gregaria. J Insect Physiol 27: 629-639.

Dow JAT Gupta BL Hall TA and Harvey WR (1983) X-ray microanalysis of elements in frozen-hydrated sections of an electrogenic K^+ transport system: the posterior midgut of tobacco hornworm (Manduca sexta) in vivo and in vitro. J Membr Biol (in press).

Durbin RP and Hill A (1981) Osmosis in epithelial membranes. J Membr Biol 61: 41-42.

Fischbarg J and Montoreano R (1982) Osmotic permeabilities across corneal endothelium and antidiuretic hormone-stimulated toad urinary bladder structures. Biochim Biophys Acta 690: 207-214.

Frizzell RA and Duffey ME (1980) Chloride activities in epithelia. Fed Proc Fed Am Soc Exp Biol 38: 144-153.

Gupta BL (1976) Water movement in cells and tissues. In: Spencer-Davies P (ed.) Perspectives in experimental biology vol.I. Pergamon, Oxford, p.25.

Gupta BL and Berridge MJ (1966) Fine structural organization of the rectum in the blowfly, Calliphora erythrocephala (Meig.) with special reference to connective tissue, tracheae and neurocecretory innervation in the rectal papillae. J Morph 120: 23-82.

Gupta BL and Hall TA (1979) Quantitative electron probe x-ray micro-analysis of electrolyte elements within epithelial tissue compartments. Fed Proc Fed Am Soc Exp Biol 38: 144-153.

Gupta BL and Hall TA (1981) Microprobe analysis of fluid transporting epithelia: evidence for local osmosis and solute recycling. In: Ussing HH et al. (1981), p.17.

Gupta BL and Hall TA (1982) Electron probe x-ray microanalysis. In: Baker PF (ed.) Techniques in cellular physiology. Elsevier/North Holland, County Clare Amsterdam, p.128/1.

Gupta BL and Hall TA (1983) Ionic distribution in dopamine-stimulated NaCl fluid-secreting cockroach salivary glands. Am J Physiol 244: R176-R186.

Gupta BL Berridge MJ Hall TA and Moreton RB (1978a) Electron probe and ion-selective microelectrode studies of fluid secretion in the salivary glands of Calliphora. J Exp Biol 72: 261-284.

Gupta BL Hall TA and Naftalin RJ (1978b) Microprobe measurements of Na, K and Cl concentration profiles in epithelial cells and intercellular spaces of rabbit ileum. Nature (London) 272: 70-73.

Gupta BL Hall TA Maddrell SHP and Moreton RB (1976) Distribution of ions in a fluid-transporting epithelium determined by electronprobe x-ray microanalysis. Nature (London) 264: 284-287.

Gupta BL Moreton RB Oschman JL and Wall BJ (1977) Transport of ions and water in animals. Academic Press, London New York San Francisco.

Gupta BL Wall BJ Oschman JL and Hall TA (1980) Direct microprobe evidence of local concentration gradients and recycling of electrolytes during fluid absorption in rectal papillae of Calliphora. J exp Biol 88: 21-47.

Hall TA and Gupta BL (1983) The localization and assay of chemical elements by microprobe methods. Q Rev Biophys 16: 279-339.

Harvey WR (1982) Membrane physiology of insects. In: Podesta RP and Timmers SF (eds.) Membrane physiology of invertebrates. Marcel Dekker, New York, p.495.

Hill AE (1975) Solute-solvent coupling in epithelia: a critical examination of the standing-gradient osmotic flow theory. Proc R Soc London B190: 99-114.

Hill S and Hill AE (1978) Fluid transfer by Necturus gall bladder epithelium as a function of osmolarity. Proc R Soc London B200: 151-162.

Kaye GI Wheeler HO Whitlock RT and Lane N (1966) Fluid transport in the rabbit gall bladder. J Cell Biol 30: 237-268.

Koefoed-Johnsen V and Ussing HH (1958) The nature of the frog skin potential. Acta Physiol Scand. 42: 298-308.

Komnick H Schmitz M and Hinssen H (1980) Biochemical demonstration of HCO_3-und Cl-dependent ATPase activities in the rectum of larval dragonflies and inhibition of rectal chloride uptake by thiocyanate. Euro J Cell Biol 20: 217-227.

Lubbock R Gupta BL and Hall TA (1981) Novel role of calcium in exocytosis: mechanism of nematocyst discharge as shown by x-ray microanalysis. Proc Natl Acad Sci USA 78: 3624-3628.

Machen TE and Diamond JM (1969) An estimate of the salt concentration in the lateral intercellular spaces of rabbit gallbladder during maximal fluid transport. J Membr Biol 1: 194-213.

Macknight ADC and Leaf A (1978) The sodium transport pool. Am J Physiol 234: F1-F9.

Maddox J (1983) Is biology now part of physics? Nature (London) 306: 311.

Marshall AT (1980) Quantitative x-ray microanalysis of frozen-hydrated bulk biological specimens. SEM/1980/II: 335-348.

Matteucci C and Cima A (1845) Memoire sur l'endosmose. Ann Chim Phys 13: 63-86.

Persson BE and Spring KR (1982) Gall bladder epithelial cell hydraulic water permeability and volume regulation. J Gen Physiol. 79: 481-505.

Rick R Dorge A Arnim E and Thurau K (1978a) Electron microprobe analysis of frog skin epithelium. J Membr Biol 39: 313-331.

Rick R Dorge A Macknight ADC Leaf A and Thurau K (1978b) Electron microprobe anaylsis of the different epithelial cells of toad urinary bladder. J Membr Biol 39: 257-271.

Rick R Roloff C Dorge A Beck FX and Thurau K (1983) Intracellular electrolyte concentrations in the frog skin epithelium: effect of vasopressin and dependence on the Na concentration in the bathing media. J Membr Biol (in press).

Sackin H and Boulpaep EL (1975) Models for coupling of salt and water transport. J Gen Physiol 66: 671-733.

Simon M Curci S Gobler B and Fromter E (1981) Attempts to determine the ion concentrations in the lateral spaces between cells of Necturus gallbladder epithelium with microelectrodes. In: Ussing HH et al. (eds.) Water transport across epithlia. Munksgaard, Copenhagen, p.52.

Skou JC (1957) The influence of some cations on the adenosine triphosphatase from peripheral nerve. Biochim Biophys Acta 23: 394-401.

Smithson KW Millar DB Jacobs LR and Gray GM (1981) Intestinal diffusion barrier: unstirred water layer or membrane surface mucous coat? Science (Washington) 214: 1241-1243.

Spring KR and Hope A (1979) Size and shape of the lateral intercellular spaces in a living epithelium. Science (Washington) 200: 54-58.

Townson R (1799) Tracts and observations in natural history and physiology. J White, London.

Ussing HH and Zerahn K (1951) Active transport of sodium as the source of the electrical current in the short-circuited isolated frog skin. Acta Physiol Scand 23: 110-127.

Ussing HH Bindslev N Lassen NA and Sten-Knudsen (1981) Water transport across epithelia. Munksgaard, Copenhagen.

Wall BJ Oschman JL and Schmidt-Nielsen B (1970) Fluid transport: Concentration of the intercellular compartment. Science (Washington) 167: 1497-1498.

Zeuthen T (1983) Ion activities in the lateral intercellular spaces of gallbladder epithelium transporting at low external osmolarities. J Membr Biol 76: 113-122.

Zeuthen T and Monge C (1975) Intra- and extracellular gradients of electrical potential and ion activities of the epithelial cells of the rabbit ileum in vivo recorded with microelectrodes. Phil Trans R Soc London B71: 277-281.

Subject Index

A

ACTH, 74
Active transport, *see* transport
Acetazolamide, 42, 43, 54, 101
Acetylcholine, 58
Adenine, 148
Adenosine, 148
Adenylate, 25, 146, 148, 150, 152
Adenylate energy charge, 25,
  148, 150, 152
ADP, 25, 146
Adrenaline, 57
Aldosterone, 134-135, 199
Amiloride, 42, 54, 55, 101,
  136, 138, 163, 196
Amino-acids, 152
Ammonia, *see* NH_4^+
AMP, 25
AMP deaminase, 148
Amphibians, *see* taxonomic
  index
Amphotericin b, 55, 102
Angiotensin II, 75
Annelids, *see* Taxonomic
  index
Anoxia, 53, 147
Antidiuretic hormones, 11, 125
Apical folds, *see* infoldings
Asymmetric diffusion, 114
ATP, 24, 25, 146, 148,
  150, 152, 158-170
ATP/ADP ratio, 148, 149, 150,
  152
ATP turnover, 148
Autoradiography, *see* $(Na^+ + K^+)$
  ATPase localization

B

Behavioral osmoregulation, 119
Bicarbonate, 42, 101
Bioelectricity, 68
Birds, *see* taxonomic index
Body wall, 18
Brain, 10
Brush-border, 74

C

Carbachol, 58
Carrier-mediated transport, *see*
  transport
Catecholamines, 57
Chloride cells, 52-61, 69, 148,
  150, 181
Choline, 196
Chromatography, *see* HPLC
Cl^--flux, *see* flux
Cl^-/HCO_3^- co-transport, *see*
  transport
Cloaca, 114, 117
Cloacal absorption, 132
Cnidaria, *see* taxonomic index
Colon, 133, 136-137
Coprodeum, 133, 135-138
Corticosterone, 134-135
Cortisol, 59, 74, 97
Coupled transport, *see* transport
Crustaceans, *see* taxonomic index
Cuticle, 19
Cyclic AMP, 57, 59
Cyrtocyte, 8

Taxonomic Index_a

---

a : words as appearing in texts.